Thomas Allen Britton

A Treatise on the Origin, Progress, Prevention, and Cure of Dry Rot

in Timber

Thomas Allen Britton

A Treatise on the Origin, Progress, Prevention, and Cure of Dry Rot in Timber

ISBN/EAN: 9783337141509

Printed in Europe, USA, Canada, Australia, Japan

Cover: Foto ©Lupo / pixelio.de

More available books at **www.hansebooks.com**

A TREATISE

ON THE

ORIGIN, PROGRESS, PREVENTION, AND CURE

OF

DRY ROT IN TIMBER.

WITH REMARKS ON

THE MEANS OF PRESERVING WOOD FROM DESTRUCTION BY SEA WORMS, BEETLES, ANTS, ETC.

BY

THOMAS ALLEN BRITTON,

LATE SURVEYOR TO THE METROPOLITAN BOARD OF WORKS, AND SILVER MEDALLIST OF THE ROYAL INSTITUTE OF BRITISH ARCHITECTS IN 1854, 1856, AND 1870.

LONDON:
E. & F. N. SPON, 48, CHARING CROSS.
NEW YORK: 446, BROOME STREET.
1875.

THIS VOLUME

IS

𝔇𝔢𝔡𝔦𝔠𝔞𝔱𝔢𝔡 𝔱𝔬

GEORGE VULLIAMY, Esq.,

VICE-PRESIDENT

OF THE

ROYAL INSTITUTE OF BRITISH ARCHITECTS;

AND

ARCHITECT

OF

THE METROPOLITAN BOARD OF WORKS;

AS A SLIGHT ACKNOWLEDGMENT

OF HIS

COUNSEL, SYMPATHY, AND FRIENDSHIP,

DURING MANY YEARS.

PREFACE.

IN preparing this treatise on Dry Rot, the author has endeavoured to place in as condensed a form as was consistent with the nature of the subject, the knowledge and information dispersed through a numerous collection of writers who have treated thereon; he has also availed himself of the assistance of professional friends, builders, timber-merchants, foremen and carpenters; and, by so doing, has been enabled to record several instances of the progress and cure of dry rot. He has consulted many valuable papers published during the last thirty years, in the various professional journals in England, America, France, and Germany, upon this important subject, and has also obtained much useful information from the works of Evelyn, Nicholson, Tredgold by Hurst, Papworth, Burnell, Blenkarn, and other English writers upon timber; Silloway, of North America; Porcher, of South America; Du Hamel, De Morny, and De Lapparent, of France; and several writers whose works will be referred to.

It is many years since a separate and complete work on dry rot has been published, and those who are desirous of inquiring into the matter are frequently at a loss where to obtain any information. Existing works on the subject are out of print, and although they can be seen at a few professional institutes, they are beyond the reach of the general public.

It has been the aim of the author in preparing this treatise to give a fair hearing to every patentee, and he has endeavoured to be as impartial as possible in recording instances of failure and success. If he has erred in any particular case, he will be happy, should this work reach a second edition, to make any necessary correction.

The reader will probably find some things repeated in the course of the work; this is in many cases unavoidable, and in some advisable; for if by a little tautology important truths can be impressed upon the mind of the reader, the author will feel that his labour in preparing this work has not been altogether in vain.

Modern authorities have been relied upon in preference to ancient ones: the following sentence, written by the late Sydney Smith, is quoted as a reason for so doing:

"Those who come first (our ancestors) are the young people, and have the least experience. We have added to their experience the experience of many centuries; and, therefore, as far as experience goes, are wiser, and more capable of forming an opinion than they were."

20, Limes Grove, Lewisham,
May 14*th*, 1875.

CONTENTS.

CHAPTER I.
On the Nature and Properties of Timber .. .. Page 1

CHAPTER II.
On the Gradual Rise and Development of Dry Rot 14

CHAPTER III.
On Felling Timber 51

CHAPTER IV.
On Seasoning Timber by Natural Methods, viz. Hot and Cold Air; Fresh and Salt Water; Vapour; Smoke; Steam; Boiling; Charring and Scorching, &c. 63

CHAPTER V.
On Seasoning Timber by Patent Processes, &c. 105

CHAPTER VI.
On the Means of Preventing Dry Rot in Modern Houses 171

CHAPTER VII.
On the Means of Preservation of Wooden Bridges, Jetties, Piles, Harbour Works, &c., from the Ravages of the *Teredo navalis* and other Sea-worms 203

CHAPTER VIII.

On the Destruction of Woodwork in Hot Climates by the *Termite* or White Ant, Woodcutter, Carpenter Bee, &c.; and the Means of Preventing the Same .. .. Page 240

CHAPTER IX.

On the Causes of Decay in Furniture, Wood Carvings, &c.; and the Means of Preventing and Remedying the Effects of such Decay 262

CHAPTER X.

Summary of Curative Processes 283

CHAPTER XI.

General Remarks and Conclusion 288

INDEX 295

ILLUSTRATIONS.

DRY ROT ON FLOOR JOIST	*Frontispiece*
	To face page
TIMBER BEAMS—ROTTEN AT THE HEART	34
BALTIC MODES OF CUTTING DEALS	64
MR. KYAN'S TIMBER PRESERVING TANK	126
MESSRS. BETHELL AND CO.'S TIMBER PRESERVING APPARATUS..	136
TIMBER PILES FROM BALACLAVA HARBOUR	208
DESTRUCTION OF TIMBER PILE BY TEREDO..	212
SHELL AND CELL OF TEREDO NAVALIS	216
PILES, SOUTHEND PIER; LIMNORIA, &C.	220
CARPENTER BEES AT WORK	260

A TREATISE

ON

DRY ROT IN TIMBER.

CHAPTER I.

ON THE NATURE AND PROPERTIES OF TIMBER.

In considering the subject of Timber trees, we commence with their Elementary Tissues, and first in order is the *Formative Fluid,* which is the sole cause of production of every tissue found in trees. It is semi-fluid, and semi-transparent, and in this condition is found abundantly between the bark and the wood of all trees in early spring; and thus separates those parts so as to permit the bundles of young wood to pass down from the leaves, and thus enable the tree to grow. It is under these conditions that the woodman strips the bark from trees which are to be cut down, since then it does not adhere to the wood.

The first step in the formation of any tissue from the formative fluid is the production of a solid structureless fabric called *Elementary Membrane,* and a modification of that fabric termed *Elementary Fibre.*

The structures which are produced from the above-mentioned "raw material" are very varied in appearance, and are called *Cellular Tissues*, to signify that they are made up of hollow cells. The spaces between the cells are called *Intercellular Spaces*, which are of vital importance, as they contain air. *Woody fibre* constitutes the mass of the stems of our forest trees. Its peculiar characteristic is that of great tenacity, and power of resistance, and for this its structure is admirably adapted: it consists of bundles of very narrow fibres, with tapering extremities, and is so placed from end to end, that the pointed ends overlap each other. Each fibre is very short, and the partitions which result from the apposition of the fibres, end to end, do not interfere with the circulation through them. The tube is not composed of simple thin membranes only; but in addition has a deposit within it, which, without filling the tube, adds very greatly to the strength of the fibre: an arrangement whereby the greatest strength and power of resistance and elasticity shall be obtained; and, at the same time, the functions of circulation uninterruptedly maintained. The strength is mainly due to the shortness of each fibre, the connection by opposite ends of many fibres, almost in one direct line, from the root upwards; and lastly, to the deposit on the inner side of the membrane. The uses of woody fibre are very varied and most important; it is the chief organ of circulation in all wooded plants, and, for this purpose, pervades the plant from the root to the branches. The current in this tissue is directed upwards from the shoot, through the stem to the leaves, and downwards from the leaves through the bark to the root. Thus, its current has a twofold tendency; the ascending and chief one being for the purpose of taking **the raw, or** what is called the *common* **sap**, from the ground to be digested in **the** leaves, and the descending being devoted **to the removal from the leaves of the** digested, or what is **termed the** *proper sap*, **to** be applied to the purposes **of the tree, and also of the** refuse matter to be carried **to the roots, and thence** thrown out into the **soil as a noxious material. The** *proper sap* differs considerably in different trees; **it is always less** liquid, **and contains a** much **greater** proportion **of vegetable matter** than the common sap. **It is very probable that trees of** the same kind produce **proper sap of** different **qualities in** different **climates**.

Woody Fibre **may be considered the storehouse of the** perfected **secretions. It is well known that as trees advance** in life, **the wood assumes a darker colour, and more** particularly that **lying near to the centre of the** stem. This is due to **the** deposit of the perfected juices in the woody fibre **at that point; and where** age has matured the **tree,** it is probable that the woody fibre so employed is no **longer** fitted for the circulation of the sap; and, also, that **the** perfected sap, when once deposited, does not again join in the general circulation. The dark colour **of the heart of oak,** as contrasted **with oak of very recent growth, is an illustration** of this fact, **as is also** the deep colour **which is** met **with in ebony and** rosewood. **Technically, the** inner wood is **called** the heart-wood, and **the outer** or younger wood the sap-wood. Of these, **the** former contains little fluid, and no vegetable life, and, being the least liable to decay, is therefore the most perfect wood; the latter is soft and perishable in its nature, abounding in fermentable elements; thus affording the very food for worms, whose destructive inroads hasten its natural tendency to decay.

The proportion of sap-wood in different trees varies very much. Spanish chestnut has a very small proportion of sap-wood, oak has more, and fir a still larger proportion than oak; but the proportions vary according to the situation and soil, and according to the age at which they have been felled: for instance, the teak tree in Malabar, India, differs from teak in Anamalai, South India. This subject has been very fully treated by Mr. Patrick Williams, in his valuable work on Naval Timber.

WOODED STEMS are divided into two great and well-defined classes, according to their internal conformation, viz. such as grow from without (exogenous), and such as enlarge from within (endogenous). The former are more common in cold, and the latter in hot climates.

EXOGENOUS STEMS.—On examining a section of a stem of an oak, or any other of our forest trees, we observe the following parts: first, the pith, or its remains in the centre; second, the bark on the outside; third, a mass of wood between the two, broken up into portions by the concentric deposition of the layers, and by a series of lines which pass from the centre to the circumference. Thus, there are always pith, bark, wood, and medullary rays. Each stem has two systems, the cellular or horizontal, and the vascular or longitudinal, and the parts just mentioned must belong to one or other of those systems. Thus, the pith, medullary rays, and bark belong to the horizontal system; and the wood constitutes the longitudinal system.

THE PITH occupies the centre of the stem, and remains throughout the period of growth of some trees, as of the elder; or is abstracted after a few years, as in the oak, and almost all large trees. In the latter class of trees, there are some remains of the pith for many years after the process of absorption has commenced, but at length no vestige can be detected, and its position is known only by the central spot around which the wood is placed in circles. In the old age of the tree the pith frequently assumes a colour which it has obtained from the juices which have been deposited. The connections of the pith are extremely important. Firstly, it is in direct connection with every branch, and is the structure which first conveys fluids to, and receives fluids from every new leaf. It thence becomes the main organ of nutriment, and, at the same time, the chief depository of the secretions. Secondly, it is in equally direct and unbroken connection with the bark, through the medium of the medullary rays; and so becomes the centre of all the movements of sap which proceed in the horizontal system.

The mode in which the ultimate disappearance of the pith occurs has been a matter of speculation. That the circulation in the heart-wood ceases after a certain number of years, and that the connection between it and the bark becomes broken, is proved by the fact that numbers of trees may be found in tolerably vigorous growth within the bark, whereas at the heart they are decayed and rotten. It appears clear that it is not converted into wood, and there are facts against the opinion that it is gradually compressed by the wood; but since it is known that in the growth of the tree much compression of the previously formed wood must occur, and since this compression is a likely theory by which to account for the disappearance of the less resisting pith, it is now generally considered to be one of the causes of this occurrence. As a general rule, the pith, so long as it exists, is not mingled with other than cellular structures; but, in certain instances, wooden fibre has been found with it, and, in others, spiral vessels have been detected.

MEDULLARY SHEATH.—Immediately surrounding the pith of all exogenous plants, there is a layer of longitudinal tissue, which has received the name of medullary sheath. This sheath has no special walls, but is bounded by the pith on the inner, and the wood on the outer side. It is in this situation that ducts of various kinds and spiral vessels may be found, and in all cases it conveys the longitudinal structure from the root, direct to each leaf. The integrity of this structure is therefore highly necessary to the life of the tree.

MEDULLARY RAYS.—These structures come next in order, and, as has been previously intimated, belong to the horizontal cellular system of the stem; they constitute the channel of communication between the bark and the pith, and are composed of a series of walls of single cells resting upon the root, and proceeding to the top of the tree, and radiating from the centre. They lie between the wedge-like blocks of wood, and as they have a lighter colour than the wood, they are evident on an oblique section of any stem, and are called the silver grain. Their colour and number suffice to enable anyone to distinguish various kinds of wood, and greatly increase their beauty. They cannot, of course, exist before the wood is formed, and are therefore not met with in very young trees. They commence to exist with the first deposited layers of wood, and continue to grow outwardly, or nearest to the bark, so long as the wood continues to be deposited. In those woods which possess in abundance the silver grain, another source of ornament exists, viz. a peculiar damask or dappled effect, somewhat similar to that artificially produced on damask linens, moreens, silks, and other fabrics, the patterns on which result from certain masses of the threads on the face of the cloth running lengthways, and other groups crossways. This effect is observable in a remarkable degree in the more central planks of oak, especially in Dutch wainscot.

THE BARK.—As the medullary rays terminate in the bark, on their outer side, the consideration of that part next follows. It forms the sheath of the tree, and its more immediate use is that of giving protection to the wood. If bark did not exist, there would be no formative fluid, and without formative fluid there could not be any deposit of woody fibre.

WOOD.—We find wood occupying nearly the whole body of the trunk of the tree, and arranged, as a rule, in a very regular manner. On taking up any piece of wood, but more particularly the entire section of a stem, we first notice a series of circles, which increase in diameter and separate by wider intervals as we approach the bark.

In this manner the trunk is composed of numerous zones enclosed within each other. Again, in almost all trees, the medullary rays before mentioned may be observed passing in straight lines from the centre to the circumference; and, as the circle of the stem at the bark is much larger than any circle near to the centre, it follows that the medullary rays will be wider apart at the bark than at the pith. On this view of the subject it may be stated that the stem is composed of a series of wedge-shaped blocks, which have their edges meeting at the centre. The combination of these two views gives the correct idea of the arrangement of the wood, viz. a series of wedges, each divided into segments of unequal width by circular lines passing across them. From this description it must not be imagined that these various portions are detached from each other; for although the medullary rays and the circular mode of deposition both tend to a less difficult cleavage of the wood, they yet bind the parts very closely to each other.

The explanation of the occurrence of distinct zones of wood is, that each zone is the produce of one year, and that in our climate, more so than in tropical climates, the period of growth of wood ceases for many months between the seasons, and this induces a distinction in appearance between the last wood of a former, and the first wood of a succeeding year. This distinction is maintained throughout each year, and throughout a long series of years.

The enclosure of zone within zone, is owing to the mode in which the wood is produced, and the position in which it is deposited. Wood is formed by the leaves during the growing season, and passes down towards the root between the bark and the wood of the previous year; and, as the leaves more or less surround the whole stem, the new layer at length completes a zone, and perfectly encloses the wood of all former years. This is the explanation of the term exogenous, which is derived from two words signifying to grow outwardly, for the stem increases in thickness by successive layers on the outer side of the previously formed wood.

The thickness of the zone for the year is rarely equal around the whole circumference of the stem, and this is due to the lesser abundance of leaves on the branches of one side than on the other, or to the prevalence of winds, or some other physical cause, acting in that direction in opposition to the growing process. It should be observed that there is not in timber any appearance of a gradual change from alburnum to perfect wood. On the contrary, in all cases the division is most decided; one concentric layer being perfect wood, and the next in succession sap-wood.

The age of trees has been inferred, when a section of the whole stem could be examined, by counting the number of rings of wood which have been deposited around the pith. In tropical countries, however, this method cannot be always relied upon.

Woods are variable in quality according to the nature of the climate, and of the soil, as also in a considerable degree to the aspect in which they are situated. Trees grown slowly in open, dry, and exposed situations are more fine and close in their annual rings, and more substantial and durable, than those which are grown in close and shady forests, or rapidly reared in moist or sappy places, the latter being soft and broad in their rings, and very subject to decay; and their pith is not always quite in the centre, for the layers are variable also.

The waggon maker takes care to combine toughness and durability by selecting his wood from trees of second growth, or from trees of first growth that from infancy have stood alone, or far apart. If the soft wood trees have stood alone, and are very large (as is often the case with some of the pines), and most of the branches are near the top, the wood near the base of the trunk is sometimes found to be *shaky*. This defect is produced by the action of heavy winds on the top of the tree, which wrenches or twists the butt, and thus cleaves apart the fibres of the wood. If the main-top (*couronnement*, of French writers) of a tree dies while the tree is yet standing, it indicates that water has found its way into the trunk, and that the tree is in a state of decay.

The fir which grows on very dry marl, forms very narrow yearly rings; if on rich or damp marl, they are wide; and when on wet soil, they are again smaller. The common fir on moor soil, has even smaller yearly rings than if on dry sand or marl. From this it is evident that too wet or too dry a soil is not suitable for this tree.

The alder and the willow grow best on wet soil, and thrive but poorly when standing dry.

The weight of wood is of great importance, because its hardness, resistance, and its heating power, as well as other valuable properties, are all more or less depending upon it. In the first place, we must consider that even wood which has been forested very light will become heavy, when put for some time into water, but in such timbers the sap is already given to dissolution. If the fibre were the only substance in the wood, then the specific weight would depend upon the number of pores contained in its body; the pores are, however, filled with a substance such as resin, die, &c. Some years since, when the Indian railways were being formed, the native woodcutters were so well aware of the above-mentioned fact, that they used to cut down the soft and inferior woods in the forests; soak them in water for a certain time; and then endeavour to pass them to the railway contractors as sound, heavy, and good railway sleepers, and the latter, not being acquainted with the Indian woods, were, at first, often deceived.

The hardest, and heaviest woods come from the hotter climates; the only exception is the pine, which thrives considerably better, and furnishes heavier timber, when it has grown in colder regions, or upon high mountains.

Trees grown on northern slopes furnish lighter timber than if grown on southern or western. The soil has great influence upon the width of the yearly rings, and from this we are able to come to a conclusion in regard to the specific weight. In the fir and larch trees the wood is heaviest when their rings are smallest.

The difference in the strength of timber between the south and the north side is attributable to the grain being closer on the north side, as the sap does not rise in the same proportion as upon the south. In forest-grown wood the difference is almost imperceptible, as the sun cannot act upon the trunk of the tree; in open-grown timber, the difference is really perceptible. It is well known that all woods do not lose strength by being open grown, or, in other words, that the south side is not always weaker than the north; that theory only applies to the coniferæ species. In ash it is the opposite, as the south side is the strongest. In soft-wooded trees, as the acer species, the difference is not perceptible, as the annual rings, and the intervening cellular tissues, are so close akin as to render the wood so compact in its grain that there is no difference in its strength. The coniferæ species, or the pines, are the only classes of woods that are stronger on the north side than on the south: it is well known that the difference originates in the wood being more open in the grain on the south side than on the north.

An influence upon the specific weight is exercised by the resin, and the dye, which are contained in the interior of the wood. On level dry ground, or deep sandy soil, we find the fir beautifully red inside; but when we look at it on lias soil, it shows broad yearly rings, and hardly any colour at all. The larch tree, again, in such soil, develops itself well with a rich colour. The cause for these appearances must therefore rest with the chemical condition of the soil, and its effect upon the individuality of the fir: it is probably the nature of the soil that causes the difference of character between Honduras and Spanish mahogany; Honduras being full of black specks, and Spanish of minute white particles, as if it had been rubbed over with chalk. Oaks generally furnish good timber when grown slowly in dry ground, whilst those from wet soil appear comparatively spongy; similar results are obtained with other trees.

Many persons constantly employed on wood are of opinion that it becomes harder if it is worked or barked whilst green.

It is not safe to condemn timber, merely because long cracks are visible on the surface. Such openings are frequently only superficial, and do not penetrate deeply into the wood: in such cases it is very little weakened thereby. It is difficult to obtain timber of large scantling without some defects of this kind, but care should be taken to ascertain if they are of a serious nature.

Trees arrive at an age when their wood becomes ripe, and then they are fit for felling; but as upon the proper method and time for doing this, the prevention of dry rot frequently hinges, a separate chapter is devoted to this part of the subject.

CHAPTER II.

ON THE GRADUAL RISE AND DEVELOPMENT OF DRY ROT.

THE opinion generally received has drawn a line of discrimination between the decay accompanied by a vegetable spreading on the surface of the timber, and that which is effected by an animal existing within it, which decay is frequently denominated the worm in timber; but as each is equally entitled to the dreaded appellation, they might more justly be distinguished as the animal and vegetable rot.

The dry rot in timber derives its name from the effect produced, and not from the cause: it is so called in opposition to the wet rot, which is properly denominated, as this exists only in damp situations, and is applied to the decomposition which takes place in timber containing sap, and exposed to moisture: but although the dry rot is usually generated in moisture, in some cases it will flourish independent of extraneous humidity. Dry rot differs from wet rot in this respect, that the former takes place only when the wood is dead, whereas the latter may begin when the tree is standing.

Wet rots are composed of porous fibre running from the rot into the trunk of the tree. This rot is of a brown colour, and has an offensive smell. The evil is often found with white spots, the latter of watery substance: when it has yellow flames, it is very dangerous.

A large quantity of the vegetable kingdom consists of plants differing totally from the flowering plants in general structure, having no flowers and producing no seed properly so called, but propagating by means of minute cellular bodies, called *spores*. These highly organized vegetables are known to botanists as *Cryptogamia*. Fungi are plants in which the fructifying organs are so minute, that without the aid of a powerful microscope they cannot be detected. To the naked eye, the fine dust ejected from the plant is the only token of reproduction; this dust, however, is not truly seed, for the word seed supposes the existence of an embryo, and there is no such thing in the reproductive bodies of fungi. The correct terms are *spores*, when the seeds are not in a case; *sporidia* when enclosed in cases. The spores or sporidia are placed in or upon the receptacle, which is of very various forms and kinds, but how different soever these may be, it is the essential part of the fungus, and in many cases constitutes the entire plant. That portion of the receptacle in which the reproductive bodies are imbedded is called the *hymenium:* it is either external, as in the Agaric, where it forms gills; or included, as in the puff-balls. The *pileus* of fungi is the entire head of the plant, not a mere head covering.

Some naturalists have insisted upon the spontaneous production of fungi, while others maintain that they are produced by seed, which is taken up and supported in the air until a soil proper for its nourishment is presented, on which being deposited it springs up of various appearances according to the principle of the seed, and the nature of the recipient.

It is extremely difficult to give a logical definition of what constitutes a fungus. It is not always easy with a cursory observation under the microscope, to determine whether some appearances are produced by fungi, insects, or organic disease; experience is the safest guide, and until we acquire that we shall occasionally fail.

In the 'Index Fungorum Britannicorum,' 2479 species of British fungi are enumerated: any detailed account of the arrangement of this extensive family of plants, or of the character of even its principal sections would be impossible within the limits of this volume; all that can be attempted will be a general description of the fungi causing dry rot.

If dry rot shows itself in a damp closet or pantry, the inside of the china or delf lying there will be coated with a mould, or a fine powder like brickdust. This excessively fine powder is no other than unaccountable myriads of the reproductive spores or *seeds* of the fungus; they are red in colour, and are produced on the surface of the fungus in millions. Certain privileged cells on the face of the fungus are furnished each with four minute points at their apex, each four bearing a single brick-red, egg-shaped spore; so that the fruit is spread over the surface of the fungus in groups of fours. To see the form of these spores the highest powers of the microscope are required, and then they can only be viewed as transparent objects. If these excessively minute bodies be allowed to fall on wet flannel, damp blotting-paper, or wet wood, they immediately germinate and proceed to reproduce the parent fungus. The red skin of the spores cracks at both ends, and fine mycelial filaments are sent out: this is the "mould," spawn, or mycelium from which the new fungus (under favourable conditions of continued moisture) appears.

It matters little where we go: everywhere we are surrounded with life. The air is crowded with birds and insects; the waters are peopled with innumerable forms, and even the rocks are blackened with countless mussels and barnacles. If we pluck a flower, in its bosom we see many a charming insect. If we pick up a fallen leaf, there is probably the trace of an insect larvæ hidden in its tissue. The drop of dew upon this leaf will probably contain its animals, visible under the microscope. The very mould which covers our cheese, our bread, our jam, or our ink, and disfigures our damp walls, is nothing but a collection of plants.

The starting point of life is a single cell—that is to say a microscopic sac filled with liquid and granules, and having within it a nucleus, or smaller sac. From this starting point of a single cell, this is the course taken: the cell divides itself into two, the two become four, the four eight, and so on, till a mass of cells is formed.

The researches of Pasteur show that atmospheric dust is filled with minute germs of various species of animals and plants, ready to develop as soon as they fall into a congenial locality. He concludes that all fermentation is caused by the germination of such infinitesimal spores.

That they elude observation does not seem strange, when we consider that some infusoria are only $\frac{1}{240000}$ of an inch in length.

It is ascertained that fungi produce seed which contains the properties of germination; and that vegetable corruption is suited to effect it. When we contemplate the fineness and volatility of the germs, the hypothesis will not appear unreasonable that they are conveyed by the rains into the earth, and are absorbed by vegetables; that with the sap they are disseminated throughout the whole body, and begin to germinate as soon as the vegetable has proceeded to corruption. Whatever, therefore, may be the appearance or situation of the fungus producing the dry rot, or from whatever substance it originates, that substance must be in a corrupt state.

Fungi result from, or are attendant on, vegetable corruption, assisted by an adequate proportion of heat and moisture. The sap, or principle of vegetation, brought into activity, is, according to the 'Quarterly Review,' No. 15, the cause of dry rot, in as far as it is favourable to the growth of fungi, as it would seem to be when in a state of fermentation.

Vegetable corruption invariably presupposes fermentation.

Fermentation is a state of vegetable matter, the component parts of which have acquired sufficient force to produce an intestinal motion, by which the earthy saline, the oily and aqueous particles therein contained, exert their several peculiar attractive and repulsive powers, forming new combinations, which at first change, and at length altogether destroy the texture of the substance they formerly composed.

There are two things essential towards creating and supporting the intestinal motion, namely, heat and humidity; for without heat, the air, which is supposed to be the cohesive principle of all bodies, cannot be so rarefied as to resume its elasticity; and without humidity there can be no intestinal motion.

According to Baron Liebig, the decay of wood takes place in the three following modes:—First, oxygen in the atmosphere combines with the hydrogen in the fibre, and the oxygen unites with the portion of carbon of the fibre, and evaporates as carbonic acid: this process is called decomposition. Second, we have to notice the actual decay of wood which takes place when it is brought in contact with rotting substances; and the third process is called putrefaction. This is stated by Liebig to arise from the inner decomposition of the wood in itself: it loses its carbon, forms carbonic acid gas, and the fibre, under the influence of the latter, is changed into white dust.

The fungus occasioning the dry rot is of various appearances, which differ according to the situation in which it exists. In the earth, it is fibrous and perfectly white, ramifying in the form of roots; passing through substances from the external surface, it somewhat differs from that form; here it separates into innumerable small branches.

Mr. McWilliam observes, "If the fungi proceed from the slime in the fissures of the earth, they are generally very ramous, having round fibres shooting in every direction. If they arise from the roots of trees, their first appearance is something like hoar frost; but they soon assume the mushroom shape."

Hence it appears that we frequently build on spots of ground which contain the fundamental principle of the disease, and thus we are sometimes foiled in our endeavours to destroy the fungus by the admission of air. In this case the disease may be encouraged by the application of air as a remedy. When workmen are employed in buildings which contain dry rot, and when they are working on ground which contains the symptoms of this disease, their health is often affected. A London builder informs us, that a few years since, while building some houses at Hampstead his men were never well: he afterwards ascertained that the ground was affected with rot, and that within one year after the house was erected, all the basement floor was in a state of premature decay. Sir Robert Smirke, architect, remarked in 1835, that he had noticed "there are certain situations in which dry rot prevails remarkably."

The fungus protruded in a very damp situation is fibrous, of moderate thickness, feels fleshy. From the spot whence it arises it extends equally around, wholly covering the area of a circle. This form would possibly continue in whatever situation it might vegetate, if the air had no motion, and every part of the substance on which it grew were equally supplied with a matter proper to encourage the expansion. The surface of this fungus is pursed, and of various colours, the centre is of a dusky brown, mixed with green, graduated into a red, which degenerates into yellow, and terminates in white.

One of the most formidable of the tribe of fungi is the *Merulius lachrymans* (often called *the* Dry Rot) of which the following description is given by Dr. Greville: "Whole plant generally resupinate, soft, tender, at first very light, cottony, and white. When the veins appear, they are of a fine yellow, orange, or reddish brown, forming irregular folds, most frequently so arranged as to have the appearance of pores, but never anything like tubes, *and distilling, when perfect, drops of water.*" Hence the term *lachrymans*, from *lacrymo*, Lat., I weep: the *Merulius lachrymans* is often dripping with moisture, as if weeping in regret for the havoc it has made. In the genus *Merulius*, the texture is soft and waxy, and the hymenium is disposed in porous or wavy toothed folds. Berkeley, in his 'Fungology,' gives the following description, which is similar to Dr. Greville's: "Large, fleshy but spongy, moist, ferruginous yellow, arachnoid and velvety beneath; margin tomentose, white; folds ample, porous, and gyroso-dentate." The *Merulius* is found in cellars and hollow trees, sometimes several feet in width, and is the main cause of dry rot.

Another formidable fungi, which attacks oak in ships, is the *Polyporus hybridus* (the dry rot of our oak-built vessels). It is thus described by Berkeley: "White, mycelium thick, forming a dense membrane, or creeping branched strings, hymenium breaking up into areæ, pores long, slender, minute."

From the slow progress dry rot makes in damp situations, it appears that excessive damps are inimical to the fungus, for its growth is more rapid in proportion as the situation is less damp, until arrived at that certain degree of moisture which is suited both to its production and vegetation. When further extended to dry situations, its effects are considerably more destructive to the timber on which it subsists: here **it is very fibrous, and in** part covered with a light brown membrane, **perfectly** soft and smooth. It is often of much greater magnitude, projecting from the timber in a white spongeous excrescence, on the surfaces of which a profuse humidity is frequently observed: at other times, it consists only of a fibrous and thin-coated web irregularly on the surface of the wood. Excrescences of a fungiform appearance are often **protruded** amidst those already described, **and are evidences of a very corrupt** matter peculiar to the spots whence they spring. According to the situation and matter in which they are produced, **they are dry and tough, or wet,** soft, and fleshy, sometimes arising in several fungiforms, each above the other, without any distinction of stem; and when the matter is differently corrupted, it not unfrequently generates the small acrid mushroom.

Mr. McWilliam observes, "The fungi arising from oak timbers are generally in clusters of from three to ten or twelve; while those from fir timber are mostly in single plants: and these will continue to succeed each other until the wood is quite exhausted."

Damp is not only a cause of decay, but is essential to it; while, on the other hand, absolute wet, especially at a low temperature, prevents it. In ships this has been particularly remarked, for that part of the hold of a ship which is constantly washed by the bilge-water is never affected with dry rot. Neither is that side of the planking of a ship's bottom which is next the water found in a state of decay, even when the inside is quite rotten, unless the rot has penetrated quite through the inside.

It matters little whether wet is applied to timber before or after the erection of a building. Timber cannot resist the effect of what must arise in either case; viz. heat and moisture, producing putrid fermentation; for instance, in basement stories with damp under them, dry timber is but little better than wet, for if it is dry it will soon be wet; decay will only be delayed so long as the timbers are absorbing sufficient moisture, therefore every situation that admits moisture is the destruction of timber.

In a constancy and equality of temperature timber will endure for ages. Sir Christopher Wren, in his letter to the Bishop of Rochester, inserted in Wadman's 'History of Westminster Abbey,' notices "That Venice and Amsterdam being both founded on wooden piles immersed in water, would fall if the constancy of the situation of those piles in the same element and temperature did not prevent the timber from rotting." Nothing is more destructive to woodwork than *partial leaks*, for if it be kept *always wet* or *always dry*, its duration is of long continuance. It is recorded that a pile was drawn up sound from a bridge on the Danube, that parted the Austrian and Turkish dominions, which had been under water 1500 years.

The writer of an article on the decay of wood, in the

'Encyclopædia Britannica,' 1855, observes, "If a post of wood be driven into the ground, the decay will commence at the surface of the ground; if driven into the earth through water, the decay will commence at the surface of the water; if used as a beam let into a damp wall, rot will commence just where the wood enters the wall."

Humboldt observes in his 'Cosmos,' with reference to damp and damp rooms, that anyone can ascertain whether a room is damp or not, by placing a weighed quantity of fresh lime in an open vessel in the room, and leaving it there for twenty-four hours, carefully closing the windows and doors. At the end of the twenty-four hours the lime should be reweighed, and if the increase exceeds one per cent. of the original weight, it is not safe to live in the room.

Decay of timber will arise from the effects of continued dryness or continued wetness, under certain conditions; or it may also arise from the effect of alternate dryness and moisture, or continued moisture with heat.

At one time dry rot appears to have made great havoc amongst the wooden ships of the British Navy. In the Memoirs of Pepys, who was Secretary to the Admiralty during the reigns of Charles II. and James II., reference is made to a Commission which was appointed to inquire into the state of the navy, and from which it appears that thirty ships, called new ships, "for want of proper care and attention, had toadstools growing in their holds as big as one's fists, and were in so complete a state of decay, that some of the planks had dropped from their sides."

In the 'European Magazine' for December, 1811, it is stated that, "about 1798, there was, at Woolwich, a ship in so bad a state that the deck sunk with a man's weight, and the orange and brown coloured fungi were hanging, in the shape of inverted cones, from deck to deck."

Mr. William Chapman, in his 'Preservation of Timber from Premature Decay,' &c., gives several instances of the rapid decay of the ships of the Royal Navy, about the commencement of the present century. He mentions three ships of 74 guns each, decayed in five years; three of 74 guns each, decayed in seven years; and one of 100 guns, decayed in six years. Mr. Pering, also, in his 'Brief Enquiry into the Causes of Premature Decay,' &c., says that ships of war are useless in five or six years; and he estimates the average duration to be eight years, and that the cost of the hull alone of a three-decker was nearly 100,000*l*. Mr. Pering was formerly at the dockyard, Plymouth, and therefore a good authority, if he availed himself of the opportunities of studying the subject. He has stated that he has seen fungi growing so strong betwixt the timbers in a man-of-war, as to force a plank from the ship's side half an inch.

No doubt a great deal of this decay was attributable to the use of unseasoned timber, and defective ventilation; but there is too much reason to believe that it was principally owing to the introduction of an inferior species of oak (*Quercus sessiliflora*) into the naval dockyards, where, we imagine, the distinction was not even suspected. The true old English oak (*Quercus robur*) affords a close-grained, firm, solid timber, rarely subject to rot; the other is more loose and sappy, very liable to rot, and not half so durable.

One cause of the decay of wood in ships is the use of wooden treenails. A treenail is a piece of cleft wood (made round), from 1 foot to 3 feet 6 inches in length and $1\frac{1}{2}$ inch in diameter. As the treenails are also made to drive easy, they never fill the holes they are driven into; consequently, if ever it admits water at the outer end, which, from shrinking, it is liable to do, that water immediately gets into the middle of the plank, and thereby forms a natural vehicle for the conveyance of water. The treenail is also the second thing which decays in a ship, the first, generally, being the oakum. Should any part of the plank or timbers of a ship be in an incipient state of decay, and a treenail come in contact with it, the decay immediately increases, while every treenail shares the same fate, and the natural consequence is, the ship is soon left without a fastening. Treenails in a warm country are sure to shrink and admit water.

Mr. Fincham, formerly Principal Builder in Her Majesty's dockyard, Chatham, considers that the destruction of timber by the decay commonly known as dry rot, cannot occur unless air, (?) moisture, and heat are all present, and that the entire exclusion of any of the three stays the mischief. By way of experiment, he bored a hole in one of the timbers of an old ship built of oak, whose wood was at the time perfectly sound; the admission of air, the third element, to the central part of the wood (the two others being to a certain degree present) caused the hole to be filled up in the course of twenty-four hours with mouldiness, which very speedily became so compact as to admit of being withdrawn like a stick.

The confinement of timber under most circumstances is attended with the worst consequences, yet a partial ventilation tends to fan the flame of decay.

The admission of air has long been considered the only means of destroying the fungus, but as it has frequently proved ineffectual, it must not be always taken as a certain remedy. If dry air be properly admitted, in a quantity adequate to absorb the moisture, it will necessarily exhaust and destroy the fungus; but care should be taken lest the air should be conveyed into other parts of the building, for, after disengaging itself from the fungus over which it has passed, it carries with it innumerable seeds of the disease, and destroys everything which offers a bar to its progress. Air, in passing through damps, will partake of their humidity; it therefore soon becomes inadequate to the task for which it is designed. Owing to this circumstance, air has been frequently admitted into the affected parts of a building without any ultimate success; too often, instead of injuring the fungus, it has considerably assisted its vegetation, and infected with the disease other parts of the building, which would otherwise probably have remained without injury. The timber, which is in a state of decomposition by an intestinal decay, is little affected by the application of air, as this cannot penetrate the surrounding spongeous rottenness which generally forms the exterior of such timber, and protects the action which the humid particles have acquired in the exterior: as the extent and progress of the disease is therefore necessarily concealed, it is difficult to ascertain correctly the effect produced by the admission of dry air. Under these circumstances of necessity and danger, it will require considerable skill to effect the purpose without increasing the disease, and, as each case has its own peculiar characteristics, it is necessary before one attempts to admit air as a remedy, to previously estimate the destructive consequences which may result from so doing, and ascertain whether it will be injurious or beneficial to the building. The joists of the houses built by our ancestors last almost for ever, because they are in contact with an air which is continually changing. Now, on the contrary, we foolishly enclose them between a ceiling of plaster (always very damp to begin with) and a floor; they frequently decay, and then cause the most serious disasters, of which it is impossible to be forewarned.

Damp, combined with warmth, is as a destroying agent, still more active than simple damp alone—the heat being understood as insufficient to carry off the moisture by evaporation; and the higher the temperature with a corresponding degree of moisture, the more rapid the decay. If the temperature to which wood is exposed, whilst any sap remains in it, is too elevated, the vegetable fluids ferment; the tenacity is diminished, and when the action is carried to its full extent, the wood quickly becomes affected by the dry rot. Exposure to the atmosphere in positions where rain can lodge in quantity, contact with the ground, and application in damp situations deprived of air, will render wood liable to the wet rot; and however well seasoned it may have been previously to being brought within the influence of any of these causes, it will infallibly suffer. Air should therefore have free access to the wood in every direction:

..... "for without in the wall of the house he made narrowed rests round about, that the beams should not be fastened in the walls of the house."—1 Kings vi. 6.

Rondelet says, "The woodwork of the church of St. Paul, outside the city walls, which was destroyed by fire in 1823, was erected as far back as the fifth century." Although the atmosphere surrounding the framework was often at once warm and damp, yet it was never stagnant. It should be remembered that 500 people in a church during two hours give off fifteen gallons of water into the air, which, if not carried away, saturates everything in the building after it has been breathed over and over again in conjunction with the impurities it contains collected from each individual.

Fever, scrofula, and consumption arise in many instances from defective ventilation.

The signs of decay in timber are, as has been stated, fungi. Some of them now and then are microscopic, and owe their existence to the sporules deposited on the surface; while fermentation, generated by prolonged contact with warm, damp, and stagnant air, is as a soil where seeds sow and nourish themselves.

Mr. McWilliam, in his work on dry rot, states that if the temperature be very low or very high, the effects are the same with respect to the growth of fungi. At 80° dry rot will proceed rapidly, at 90° its progress is more slow; at 100° it is slower still, and from 110° to 120° it will in general be arrested. It will proceed fast at 50°; it may be generated at 40°; its progress will be slow at 36°; and is arrested at 32°, yet it will return if the temperature is raised to 50°.

Dry rot externally first makes its appearance as a mildew, or rather a delicate white vegetation, that looks like such. The next step is a collecting together of the fibres of the vegetation into a more decided form, somewhat like hoar frost; after which it speedily assumes the leathery, compact character of the fungus, forming into leaves, spreading rapidly in all directions, and over all materials, and frequently ascending the walls to a considerable height, the colour variable—white, greyish white, and violet, light or decided brown, &c.

In the section of a piece of wood attacked by dry rot a microscope reveals minute white threads spreading and ramifying throughout its substance; these interlace and become matted together into a white cottony texture, resembling lint, which effuses itself over the surface of the timber; then in the centre of each considerable mass a gelatinous substance forms, which becomes gradually of a yellow, tawny hue, and a wrinkled, sinuated porous consistence, shedding a red powder (the spores) upon a white down; this is the resupinate pileus, the hymenium being upwards, of *Merulius lachrymans*, in its perfect and matured state. Long before it attains to this, the whole interior of the wood on which it is situated has perished; the sap vessels being gradually filled by the cottony filaments of the fungus; no sooner do these appear externally than examination proves that the apparently solid beam may be crumbled to dust between the fingers; tenacity and weight are annihilated.

Dr. Haller says that seven parts in eight of a fungus in full vegetation are found by analysis to be completely aqueous.

The strength of fungi is proportionate to the strength of the timber the cohesive powers and nutritive juices of which they absorb; and according to the food they receive so they are varied and modified in different ways, and are not always alike. Different stages of corruption produce food of different qualities, and hence many of the different appearances of fungi. One takes the process of corruption up where another leaves it off, and carries it forward and farther forward to positive putrefaction.

The forms which fungi assume are extremely diversified; in some instances we have a distinct stem supporting a cap, and looking somewhat like a parasol; in others the stem is entirely absent, and the cap is attached either by its margin, and is said to be *dimidiate*, or by its back, or that which is more commonly its upper surface, when it is called *resupinate*. In some species the form is that of a cup, in others of a goblet, a saucer, an ear, a bird's nest, a horn, a bunch of coral, a ball, a button, a rosette, a lump of jelly, or a piece of velvet.

Decomposition takes place without fungus where the timber and the situation are always moist, as in a close-boarded kitchen floor, where it is always dry, or very nearly so, and where it is alternately wet and dry, cold and hot. When the decomposition is affected with very little moisture, and no fungus, the admission of air will generally prevent further contamination; but where there is abundance of moisture, rottenness, and fungus, a small quantity of air will hasten the destruction of the building.

In timber which has been only superficially seasoned this disease is produced internally, and has been known to convert the entire substance of a beam, excepting only the external inch or two of thickness to which the seasoning had penetrated, into a fine, white, and thread-like vegetation, uniting in a thick fungous coat at the ends, the semblance being that of a perfectly sound beam. In this internal rot a spongy fungous substance is formed between the fibres. This has often been observed in large girders of yellow fir, which have appeared sound on the outside, but by removing some of the binding joists have been found completely rotten at the heart. An instance of this kind occurred at Kenwood (the seat of the Earl of Mansfield) in 1815. Major Jones, R.E., states that on one occasion he was called upon to report on the state of a building in Malta; that the timbers had every external appearance of being sound, but on being bored with an auger they were found internally in a total state of decay. It is on this account that the practice of sawing and bolting beams is recommended, for when timber is large enough to be laid open in the centre this part is laid open to season; so that when a tree is large enough to be cut through to make two or more beams, decomposition is impeded.

The first symptoms of rottenness in timber are swelling,

discoloration, and mouldiness, accompanied with a musty smell; in its greater advance the fibres are found to shrink lengthways and break, presenting many deep fissures across the wood; the fibres crumble readily to a fine snuff-like powder, but retain, when undisturbed, much of their natural appearance.

In whatever way boughs are removed from trees, the effect of their removal is, however, very frequently to produce a rotting of the inner wood, which indicates itself externally by a sudden abnormal swelling of the trunk a little above the root; sometimes the trunk becomes hollow at the part affected, and this particular description of rot will almost invariably be found to exist in those trees whose roots are much exposed. The rot itself is either of a red, black, or white colour in the timber when felled, and when either of the two last-named colours prevail, it will be found that the decay does not extend very far into the tree; but if, on the contrary, the colour of the parts most visibly affected should be decidedly red, the wood should be rejected for any building purposes. Sometimes small brown spots, indicative of a commencement of decay, may be observed near the butt or root end of trees, and though they do not appear to be connected with any serious immediate danger to the durability of the wood, it is advisable to employ the material so affected only in positions where it would not be confined in anything like a close, damp atmosphere.

Great hesitation may be admitted as to the use of timber which presents large bands of what are supposed

D

to be indefinitely-marked annual growth, because the existence of zones of wood so affected may be considered to indicate that the tree was not in a healthy state when they were formed, and that the wood then secreted lacked some of the elements required for its durability, upon being subsequently exposed to the ordinary causes of decay.

In many cases when timber trees are cut down and converted for use, it is found that at the junction of some of the minor branches with the main stem, the roots, as it were, of the branches traverse the surface wood in the form of knots, and that they often assume a commencement of decay, which in the course of time will extend to the wood around them. This decay seems to have arisen in the majority of cases from the sudden disruption of the branch close to its roots, with an irregular fracture, and with such depressions below the surface as to allow the sap to accumulate, or atmospheric moisture to lodge in them. A decomposition of the sap takes place—in fact, a wound is made in the tree—and what are called "druxy knots" are thus formed, which have a contagious action on the healthy wood near them.

There is this particular danger about the dry rot; viz. that the germs of the fungi producing it are carried easily, and in all directions, in a building wherein it once displays itself, without necessity for actual contact between the affected or the sound wood; whereas the communication of the disease resulting from the putrefactive fermentation, or the wet rot, only takes place by actual contact.

Before dry rot has time to destroy the principal timbers in a building, it penetrates behind the skirtings, dadoes,

Timber Beams, — rotten at the heart

and wainscotings, drawing in the edges of the boards, and splitting them both horizontally and vertically. When the fungus is taken off, they exhibit an appearance similar both in back and front to wood which has been charred; a light pressure with the hand will break them asunder, even though affected with the rot but a short time; and in taking down the wainscot, the fibrous and thin-coated fungus will generally be seen closely attached to the decayed wood. In timber of moderate length the fungus becomes larger and more destructive, in consequence of the matter congenial to its growth affording a more plentiful supply.

It is a great characteristic of fungi in general that they are very rapid in growth, and rapid in decay. In a night a puff-ball will grow prodigiously, and in the same short period a mass of paste may be covered with mould. In a few hours a gelatinous mass of *Reticularia* will pass into a bladder of dust, or a *Coprinus* will be dripping into decay. Many instances have been recorded of the rapidity of growth in fungi; it may also be accepted as an axiom that they are in many instances equally as rapid in decay.

In considering the liability of any particular description of foreign timber to take the dry rot, attention must be paid to the circumstances under which it is imported. Sometimes the timber is a long while coming here, whilst at other times it is imported in a very short period. The length of time consumed in the voyage has a great deal to do with its likelihood of taking the rot: it may have a very favourable passage, or a very wet one, and the ship is frequently, in some degree, affected with the disease. It perhaps begins in the ship, and it may often be seen between the timber or deals, when it will impregnate the wood to a great depth. Whether it is inherent in the timber or not, of this we may be certain, that where there is a fetid atmosphere it is sure to grow. Canadian yellow wood pine timber is more subject to rot than Baltic or Canadian red wood timber, although the latter will sometimes decay in four or five years. Turpentine is a preventive against dry rot, and Canadian timber is sometimes largely impregnated with it, especially the red wood timber; the yellow wood is very subject to dry rot. Very few cargoes of timber in the log arrive from Canada in which in one part or other of nearly every log you will not see a beginning of the vegetation of the rot. Sometimes it will show itself only by a few reddish, discoloured spots, which, when scratched by the finger nail to the extent of each spot, it will be seen that the texture of the timber to some little depth is destroyed, and will be reduced to powder; and on these spots a white fibre may generally be seen growing. If the timber has been shipped in a dry condition, and the voyage has been a short one, there may be a few logs without a spot; but generally speaking very few cargoes arrive from Canada in which there are many logs of timber not affected. But if the cargo has been shipped in a wet condition, and the voyage has been a long one, then a white fibre will be seen growing over nearly every part of the surface of every log; and in cargoes that have been so shipped, all the logs of yellow pine, red pine, and of oak, are generally more or less affected on the surface.

Nearly every deal of yellow pine that has been shipped in Canada in a wet state, when it arrives here is also covered over with a network of little white fibres, which are the dry rot in its incipient state. There is no cargo, even that which is shipped in tolerably dry condition, in which, upon its arriving here, may not be found some deals, with the fungus beginning to vegetate on their surface. If they are deals that have been floated down the rivers of America or Canada, and shipped in a wet state, on their arrival here they are so covered with this network of the fungus, that force is often necessary to separate one deal from another, so strongly does the fungus occasion them to adhere. They grow together again, as it were, after quitting the ship, while lying in the barges, before being landed. Accordingly, if a cargo has arrived in a wet condition, or late in the year, or if the rain falls on the deals before they are landed, and they are then piled in the way in which Norwegian and Swedish deals are piled, that is, flatways, in six months time, or even less, the whole pile of deals become deeply affected with rot; so that, whenever a flat surface of one deal is upon the flat surface of another, the rot penetrates to the depth of $\frac{1}{8}$ of an inch. Its progress is then arrested by repiling the deals during very dry weather, and by sweeping the surface of each deal before it is repiled: but the best way is to pile the deals in the first instance upon their edges; by which means the air circulates freely around them, the growth of the fungus is arrested, and the necessity of repiling them prevented. If the ship is built of good, sound, and well-seasoned oak, the rot would perhaps not affect it, but in order to prevent its doing so, the precaution is usually taken to scrape the surface as soon as the hold is clear of the cargo of timber. Were the cargo not cleared, and the hold not ventilated, a ship that was permanently exposed to this fungus would, no doubt, be affected. It is easy, however, to prevent its extending by washing the hold with any desiccating solution.

Anyone who wishes to know how timber is occasionally shipped to this country should read the report of a trial, in the 'Times,' 22nd Feb., 1875 (Harrison *v.* Willis), relative to a cargo of pitch pine shipped from Sapelo, in the Isthmus of Darien, for Liverpool. This cargo, however, never arrived at Liverpool: it was lost at sea.

The motto of the Worshipful Company of Shipwrights is, "Within the ark, safe for ever." We suggest it should be altered to, "Within the ark *which is free from dry rot*, safe for ever."

There are two descriptions of European deals very liable to take the dry rot; viz. yellow Petersburgh deals, and yellow and white battens, from Dram, in Norway. When Dram battens, which have been lying a long time in bond in this country, have not been repiled in time, they have been found as much affected with the dry rot as many Canadian deals; though this has not happened in so short a time as has been sufficient to rot Canadian deals. The fungus growing on the Petersburgh deals and Dram battens has all the characteristics and effects of dry rot as exhibited in the Canadian deals, the detection of dry rot being in most cases the same.

It should be remembered that white deal absorbs more water than yellow; and yellow more water than red; and the quantity of water absorbed by the white accounts for its more rapid decay in external situations; as the greater the quantity of water absorbed the quicker is the timber destroyed. Mr. John Lingard, in his work on timber (1842), states that he has proved that $4\frac{1}{2}$ oz. of water can be driven off from a small piece of fir, weighing only 10 oz. when wet, which is nearly half. This timber was on a saw-pit, and going to be put into a building.

The most general, and the most fatal cause of decay, viz. the wet rot, has attracted less attention than the more startling, but less common evils, the dry rot, and the destruction by insects.

Sir Thomas Deane, in 1849, related before the Institution of Civil Engineers of Ireland, an extraordinary instance of the rapid decay of timber from rot, which occurred in the church of the Holy Trinity at Cork.

On opening the floors under the pews, a most extraordinary appearance presented itself. There were flat fungi of immense size and thickness, some so large as almost to occupy a space equal to the size of a pew, and from 1 to 3 inches thick. In other places fungi appeared, growing with the ordinary dry rot, some of an unusual shape, in form like a convolvulus, with stems of from a quarter to half an inch in diameter. When first exposed, the whole was of a beautiful buff colour, and emitted the usual smell of the dry-rot fungus.

During a great part of the time occupied in the repairs of the church, the weather was very rainy. The arches of the vaults having been turned before the roof was slated, the rain water saturated the partly decayed oak beams. The flooring and joists, composed of fresh timber, were laid on the vaulting before it was dry, coming in contact at the same time with the old oak timber, which was abundantly supplied with the seeds of decay, stimulated by moisture, the bad atmosphere of an ill-contrived burial-place, and afterwards by heat from the stoves constantly in use. All these circumstances account satisfactorily for the extraordinary and rapid growth of the fungi.

Many instances might be mentioned of English oak being affected with dry rot, under particular circumstances. There was a great deal at the Duke of Devonshire's, at Chiswick, about 60 years ago. Needy builders, who work for contract, sometimes use American oak, and call it wainscot: it is a bad substitute for wainscot, being very liable to warp and to be affected with dry rot. "I know of one public building," observed the late Mr. Henry Warburton, M.P., "in which it has been introduced, and, I suppose, paid for under that name."

Another serious instance of the decay of timber from rot occurred some time since in Old St. Pancras Church, London. When the dry rot made its appearance, it spread with amazing rapidity. Sometimes in the course of a night, a fungus of about the consistence of newly-fallen snow, and of a yellowish-white unwholesome colour, would be found to have spread over a considerable surface. The fungus was without shape, but in some cases it rose to a height of 2, 3, or 4 inches above the planks or other surfaces on which it grew. It could be cut with a knife, leaving a clear edge on each side, and there did not seem to be any covering or membrane over the outer or under surface. The smell of those matters was unpleasant, and seemed like the concentration of the smell which had pervaded the church for so long a time before; and, in a short time, beams, planks of flooring, railings, &c., were reduced to rottenness: the colour changed, and a heavy dark-brown dust fell, and represented the once solid timber. On making an examination with a view of discovering the cause of the attack, it was found that in the graveyard, near the church, there were graves, and several vaults: there were also vaults in the inside of the church. Most of them were filled, or nearly so, with water, which had run from the overcrowded graves.

In the interior there were water-logged vaults, and the walls were saturated with damp. It was also seen that from want of proper spouts and drains, near the outer walls, the drip from the large pent roof had fallen into the foundations. In this situation, when the window frames were properly arranged, a drain dug round and from parts of the church, and other alterations, which should long before have been made, were completed, the dry rot vanished, and no more complaints of the foulness of the air have since been heard.

We could quote many cases of rot which have been caused from the want of proper drains and spouts. Architects should remember that the feet of Gothic collar roofs have to bear the whole weight of the roof, and unless well seasoned, and carefully protected from damp, leaks, &c., premature decay and dry rot will be sure to occur. It is surprising what injury leaks from gutters will sometimes do. In 1851, Professor T. L. Donaldson stated that "a brestsummer of American timber was used some time since at a house in London: after an expiration of three years cracks began to appear in the front wall. A friend of mine, an architect, was called in to find out the cause; and after examining different parts of the house, was almost giving up his search in despair, when he thought he would have the shop cornice removed and look at the brestsummer. He then discovered that some water had been admitted by accident, and penetrating the brestsummer, had caused it to rot, and crack the wall."

Dry rot was found in the great dome of the Bank of England, London, as originally built by Sir Robert Taylor: it also existed in the Society of Arts building, in the Adelphi, London. It was also found in the domes of the Panthéon, and Halle-au-Blé, Paris; but we hope there is no dry rot in the dome of St. Paul's Cathedral, London, which is constructed entirely of timber, covered externally with lead.

The decayed state of a barn floor attacked by rot is thus described by Mr. B. Johnson: "An oak barn floor which had been laid twelve years began to shake upon the joists, and on examination was found to be quite rotten in various parts. The planks, $2\frac{1}{2}$ inches in thickness, were nearly eaten through, except the outsides, which were glossy, and apparently without blemish. The rotten wood was partly in the state of an impalpable powder, of a snuff colour; other parts were black, and the rest clearly fungus. No earth was near the wood." This oak was probably of the *Quercus sessiliflora* species ; and there was no ventilation to the floor.

Mr. John Armstrong, carpenter, employed for many years at Windsor Castle, observed: "I was employed a few years back at a house where I found a floor rotten. We took it up; it was yellow pine; it was laid in the damp, but on sleepers, and the sleepers were not rotten: they were of a different description of wood." Probably the sleepers were of Baltic red wood.

Dr. Carpenter relates an instance of the expansive power resulting from the rapid growth of the soft cellular tissue of fungi. About the commencement of this century the town of Basingstoke was paved; and not many months afterwards the pavement was observed to exhibit an unevenness which could not easily be accounted for. In a short time after, the mystery was explained, for some of the heaviest stones were completely lifted out of their beds by the growth of large toadstools beneath them. One of these stones measured 22 inches by 21 inches, and weighed 83 lb., and the resistance afforded by the mortar which held it in its place would probably be even a greater obstacle than the weight. A similar incident came under the notice of Mr. M. C. Cooke (the author of 'British Fungi'), of a large kitchen hearthstone which was forced up from its bed by an under-growing fungus, and had to be relaid two or three times, until at last it reposed in peace, the old bed having been removed to the depth of 6 inches, and a new foundation laid. A circumstance recorded by Sir Joseph Banks is still more extraordinary, of a cask of wine which, having been confined for three years in a cellar, was, at the termination of that period, found to have leaked from the cask, and vegetated in the form of immense fungi, which had filled the cellar, and borne upwards the empty wine cask.

Timber decay in contact with stone is a subject deserving consideration. This decay is entirely obviated by inserting the wood in an iron shoe, or by placing a thin piece of iron between the wood and the stone. It is said that a hard crust is formed on the timber in contact with the iron, which seems effectually to preserve it; it is, of course, necessary that a free circulation of air round the ends of the timber be provided. The most notable instance of timber decay in contact with stone with which we are acquainted occurred at the coronation of George IV. Westminster Hall was then fitted up, and they began by laying sleepers of yellow pine. The coronation was suspended for twelve months, and when the sleepers were taken up from the floor of Westminster Hall, they were in a rotten state.

Timber in contact with brickwork is in Suffolk and in some parts of England covered with sheet lead to preserve it from the effects of the damp mortar. Fungi will arise in mortar, if made with road-drift, and water from stagnant ponds, &c., and it may be traced through the mortar joints, and will thus appear on both sides of a wall. Mortar composed of unwashed sand will generate fungi; sea sand, even if washed, should never be used. It is considered that the system of grouting contributes to the early decay of timber; wood bond timber for walls has been consequently replaced by hoop iron bond. In

Manchester wood bond is frequently used, and is said to answer well, but the high temperature of the buildings may be a preventive against the decay of timber, as the walls are soon dried. The practice is a bad one.

When timber used as posts inserted in the ground is placed in the inverted position to that in which it stood when growing, it is said to be very much more durable than if placed in its natural or growing position. This is easily accounted for in the valves of the sap vessels of the growing timber opening upwards; but when that position is inverted, the valves of the sap vessels become reversed in their action; and, therefore, when timber is used as posts inserted in the ground, the valves being so reversed prevent the ascent of moisture from the soil in the wood. Mr. W. Howe relates an experiment made to test the comparative durability of posts set as they grew. He says, "Sixteen years ago I set six pairs of bar posts all split out of the butt end of the same white oak log. One pair I set butts down; another pair, one butt down, the other top down; the others top down. Four years ago those set butt down were all rotted off, and had to be replaced by new ones. This summer I had occasion to reset those that were set top down: I found them all sound enough to reset. My experiments have convinced me that the best way is to set them tops down." Other instances might be given in favour of placing posts in an inverted position in the ground. Posts will sometimes decay, for the following reason: The ends are often sawn off with a coarse implement and left spongy, with the longitudinal fibres shaken or broken a considerable way within the extremity of the wood. In this state the ends of the posts must be apt to absorb from the ground the moisture, which, being retained, and speedily pervading the whole internal surface, *especially if painted*, appears to cause decay.

With respect to the preservation of wooden fences, Mr. Cruikshank, of Marcassie, gives in detail various experiments from which it appears that—1st. When larch or pine wood is to be exposed to the weather, or to be put in the ground, no bark should be left on. 2nd. When posts are to be put in the ground, no earth should be put round them, but stones. 3rd. When a wooden fence is to be put up, a No. 4 or No. 5 wire should be stretched in place of, or alongside the upper rail.

Mr. G. S. Hartig, in the 'Revue Horticole,' gives the results of experiments made with great care and patience, upon woods buried in the earth. Pieces of wood of various kinds $3\frac{1}{8}$th inches square, were buried about one inch below the surface of the ground, and they decayed in the following order: the lime, American birch, alder, and the trembling-leaved poplar, in three years; the common willow, horse-chestnut, and plane, in four years; the maple, red beech, and common birch, in five years; the elm, ash, hornbeam, and Lombardy poplar, in six years; the oak, Scotch fir, Weymouth pine, and silver fir, were only decayed to the depth of half an inch in seven years; the larch, common juniper, red cedar, and arbor vitæ, at the end of the last-mentioned period remained uninjured. The duration of their respective woods greatly depends on their age and quality; specimens from young trees decaying much quicker than those from sound old trees; and, when well seasoned, they, of course, last much longer than when buried in an unseasoned state. In experiments with the woods cut into thin boards, decay proceeded in the following order: the plane, horse-chestnut, poplar, American birch, red beech, hornbeam, alder, ash, maple, silver fir, Scotch fir, elm, Weymouth pine, larch, locust oak.

Before quitting the subject of decay of timber when buried in the earth, it will not be out of place to allude to the decay of railway sleepers, taking for example those in India: English and American sleepers will be dealt with more in detail hereafter.

Dr. Cleghorn, Conservator of Forests, Madras Presidency, India, considers the decay of sleepers to arise in a great measure from the inferior description of wood used. Mr. Bryce McMaster, Resident Engineer, Salem, considers that the native wood sleepers in India have hitherto been found for the most part to fail on the Madras Railway, between 30 and 40 per cent. requiring to be renewed annually. Mr. McMaster undertook an investigation with a view of ascertaining the causes of this deterioration, and whether those causes could be overcome so as to render available the vast resources of India. Thirteen hundred sleepers of sixteen different woods were submitted to careful examination and scrutiny twice at an interval of one year. The sleepers were variously placed, both on embankments and in cuttings; in some cases they were entirely covered with ballast to a depth of 4 inches; while in others they were as much as possible uncovered, and completely so from the rails to the ends—the ballast being only raised 2 inches in the middle of the way, and sloped off so as to carry away the water under the rails. From these observations it appeared that only five woods, Chella wungé, Kara mardá, Palai, Karúvalem, and Ilupé, were sound at the end of two years, the other eleven not lasting even that time. Also, that when the sleepers were uncovered, decay was less rapid than when they were buried in the ballast. The plan of leaving the sleepers partially uncovered had many advantages; it effected a saving of the ballast, allowed the defects to be more quickly detected, and kept the sleepers drier. It had been urged that the heat of the sun would split the sleepers and cause the keys and treenails to shrink'; but from experience it was found that while among the "uncovered" sleepers there was a large proportion "beginning to split," or "useless from being split," there was on the other hand, among the "covered" sleepers, a still larger proportion "beginning to rot," or "useless from being rotten." It was also noticed that of the sleepers "beginning to rot," 19 per cent. had commenced under one or both chairs. This was due to the retention of moisture under them, and might be remedied by tarring the seats of the chairs. As regarded the treenails where the sleepers were rotten, the treenails were invariably found to be in the same state; while, when the heads were exposed to the sun, they were not loosened by shrinking. Another objection was, that the road would be more likely to buckle and twist, but this was not found in practice to be the case. Treenails made in India cost 2*l*. 10*s*. to 4*l*. per 1000, and the woods generally used for the purpose are Vengé, Kara mardá,

Erul, Porasa, or satin wood, and Trincomalee. The three woods first named are also extensively employed for keys, but teak keys seem to be the best, and their cost does not exceed 6*l.* per 1000. From the experience of the Indian engineers it appears that Teak, Saul, Sisso, Pedowk, Kara mardá, Acha, Vengé, Chella wungé, Palai, Erul, Karúvalem, will make very good sleepers to be used plain.

The sleepers which have failed on the Madras Railway might well be divided into two classes,—those which were originally of perishable woods, and were therefore unfit for the purpose; and those which although of good wood had been cut from young trees, and not been allowed to stand until old enough. The first arose from want of experience of the nature of Indian woods: the second from the absence of a proper system of working the jungles.

The wooden sleepers on the Indian railways should be tarred under the seats of the chairs, be laid in dry ballast, and raised slightly in the middle, and sloped off so as to throw the water under the rails. About two-thirds of the Indian woods are practically useless owing to the want of proper artificial means for preserving those of a perishable nature.

The subject of the decay of wood in India and tropical climates is too extensive to be further considered here; but is of sufficient importance to demand a volume to itself; the renewal of decayed wooden sleepers to railways forming annually a most important item in foreign railway budgets.

We have heard that some of our fortifications which have been erected within the last few years to protect our

English coast from invasion, have already been invaded by dry rot. If this be true, some one well acquainted with the subject should at once be appointed to find out the cause, and recommend the remedy in each case.

Professional men, if they wish their works to "live for ever," should consider the after consequences of neglecting to provide against dry rot. If the fungi could speak from under floors, ceiled-up roofs, behind wainscots, girders, &c., we should often hear them exclaim, "A nice moist piece of wood! Surely this belongs to us." On the beams of a building at Crawley, a carpenter many years ago cut a few words; they are full of meaning in connection with our subject, and they run as follows:

"Man of weal, beware; beware before of what cometh behind."

CHAPTER III.

FELLING TIMBER.

The end to be attained in the management of timber trees is to produce from a given number the largest possible amount of sound and durable woods. When a tree, under conditions favourable to its growth, ceases increasing the diameter of its trunk, and loses its foliage earlier in the autumn than it is wont to do, and when the top of the tree brings forth no leaves in spring, these facts may be considered as indications of decline, and that the tree is of sufficient age to be felled. The state of the upper branches of a tree may be considered to be amongst the best indications of its soundness, and provided they be in a healthy condition, the withering of the lower branches is a matter of comparatively small importance.

Trees may be considered as tall, middle rank, and low, and the size to which they will attain depends on many different circumstances. Some trees, the stems of which are short on the average, as the lime, are virtually of tall growth, from the manner in which a number of vertical branches of large size ascend from the stem. And other trees, again, whose branches are comparatively short, are of tall growth, in consequence of the length of the stems —like the beech.

The average duration of trees differs, as is well known,

in different species, and they exhibit different symptoms of decay. There are oaks in Windsor Great Park, certainly nearly one thousand years old, and which exhibit even now no appearance of approaching the end of their life. Mr. Menzies, the surveyor, in his work on Windsor Great Park, describes some of the indications of incipient decay which are peculiar to the several kinds of trees. "When a beech begins to fail," he says, "fungi appear either at the roots or on the forks, the leaves curl up as if they had been scorched, and the tree quickly perishes. In an elm, a great limb first fails, while the rest of the tree continues green and vigorous, but in a few years the whole tree suddenly dies. Coniferous trees die gradually, but quickly The oak shows the first symptoms at the points of its highest branches, while the rest of the tree will remain healthy and sound for years." This peculiarity of the oak did not escape the eye of Shakespere, that universal observer, who describes the monarch of the woods as not only having its boughs mossed with age, but its

"High top bald with dry antiquity."

The age for felling trees is a subject which calls for the deepest consideration, but does not always receive that attention which is due to its importance. Timber growers in their haste to supply the market, too often fell trees that have not arrived at maturity, the heart-wood being therefore imperfect, with much sap-wood, and, of course, little durability; but unfortunately they are the more readily led to do so on account of the increase in size being very slow after a certain age. Builders are sensible of the inferior quality of young timber in respect to duration, and it is their province to check this growing evil, by giving a better price for timber that has acquired a proper degree of density and hardness; but, unfortunately, this is an age for cheap building, without much regard being given as to durability.

Felling should not be too early, for the reasons above mentioned; neither should it be in the decline of the tree, when its elasticity and vigour are lost, and the wood becomes brittle, tainted, and discoloured, with the pith gone, and the heart in progress of decay. Maturity is the period when the sap-wood bears a small proportion, and the heart-wood has become uniform and compact. Sir John Evelyn writes, "It should be in the vigour and perfection of trees, that a felling should be celebrated." It must be obvious, however, that it is a worse fault to fell wood before it has acquired thorough firmness, than when it is just in the wane, and its heart may exhibit but the first symptoms of decay; for in the former there is no perfect enduring timber to be got, while in the latter the greater part is in the zenith of its strength.

Although there are certain symptoms by which it may be ascertained when a tree is on the decline, it is somewhat difficult to decide just when a tree is at maturity. From the investigations of naturalists, however, it may be safe to consider that hard-wood trees, as oak and chestnut, should never be cut before they are sixty years old, the average age for felling being from eighty to ninety years, and the average quantity of timber produced by a tree of that age is about a load and a half, or about 75 cubic feet.

Daviller states (see 'Cours d'Architecture') "that an oak should not be felled at a less age than sixty years." Belidor considers (see 'Sciences des Ingénieurs') "that one hundred years is the best age for the oak to be felled."

It should be remembered that the times mentioned are by no means arbitrary, for situation, soil, &c., have much to do with it. For the soft woods, as the Norway spruce and Scotch pine in Norway, the proper age is between seventy and one hundred years. The ash, larch, and elm, may be cut when the trees are between fifty and ninety years old; and between thirty and fifty years is a proper age for poplars.

The felling of timber was looked upon by ancient architects as a matter of much moment. According to Vitruvius, the proper time for felling is between October and February, and he directs that the trees should be cut to the pith, and then suffered to remain till the sap be drained out. The effusion of the sap prevents the decay of the timber, and when it is all drained out, and the wood becomes dry, the trees are to be cut down, when the wood will be excellent for use. A similar effect might be produced by placing the timber on its end as soon as it is felled, and it would, no doubt, compensate for the extra expense by its durability in use. In France, so long ago as 1669, a royal order limited the felling of naval timber from the 1st October to 15th April, when the "wind was at north," and "in the wane of the moon." Buonaparte directed that the time for felling naval timber should be "in the decrease of the moon, from 1st November to 15th March," in order to render it more durable. In England, in the first year of James I., there was an Act of Parliament prohibiting every one from cutting oak timber, except in the barking season, under a severe penalty.

James I. was not the only English sovereign who has been concerned with timber trees; for King John was obliged to cancel at Runnemede the cruel forest laws enacted by his father, William the Conqueror, especially those restricting the people from fattening their hogs.

Up to a recent period large droves of hogs were fattened upon the acorns of the New Forest in Hampshire. At the present time the hogs of Estremadura are principally fed upon the acorns of the *Ballota* oak; and to this cause is assigned the great delicacy of their flesh.

A Berkshire labourer, living near Windsor Forest, thus speaks of the delicacy of acorn-fed pork: "Well, that be pretty like the thing. I han't tasted the like o' that this many a day. It is so meller—when you gets your teeth on it, you thinks you has it; but afore you knows where you is, ain't it wanished!"

There is another point in connection with the time of felling timber, which ought to be noticed. It is a widespread opinion that trees should be felled during the wane of the moon. This planetary influence is open to doubt, but the opinion prevails wherever there are large forests. Columella, Cato, Vitruvius, and Pliny, all had their notions of cutting timber at certain ages of the moon. The wood-cutters of South America act upon it, so do their brethren in the German forests, in Brazil, and in Yucatan. It was formerly interwoven in the Forest Code of France, and, we believe, is so still. Vitruvius recommends this custom, and we find Isaac Ware writing of the suggestion: "This has been laughed at, and supposed to be an imaginary advantage. There may be good in following the practice; there can be no harm: and therefore, when I am to depend upon my timber, I will observe it." The Indian wood-cutters believe that timber is much more liable to decay, if cut when the moon is in crescent.

An American writer, in 1863, thus writes of his experience in the matter: "Tradition says that the 'old' of the moon, in February, is the best time to cut timber; but from more than twenty years of observation and actual experience, I am fully convinced it is about the worst time to cut most, if not all kinds of hard-wood timber. Birch, ash, and most or all kinds of hard wood will invariably *powder-post* if cut any time in the fall after the tree is frozen, or before it is thoroughly leaved out in the spring of the year. But if cut after the sap in the tree is used up in the growth of the tree, until freezing weather again comes, it will in no instance produce the *powder-post* worm. When the tree is frozen, and cut in this condition, the worm first commences its ravages on the inside film of the bark, and then penetrates the wood until it destroys the sap part thereof. I have found the months of August, September, and October, to be the three best in the year to cut hard-wood timber. If cut in these months, the timber is harder, more elastic, and durable than if cut in winter months. I have, by weighing timber, found that of equal quality got out for joiners' tools is much heavier when cut and got out in the above-named months than in the winter and spring months, and it is not so liable to crack. You may cut a tree in September, and another in the 'old' of the moon in February following, and let them remain, and in one year from the cutting of the first tree, you will find it sound and unhurt, while the one last cut is scarcely fit for firewood, from decay. Chestnut timber for building will last longest, provided the bark be taken off. Hemlock and pine ought to be cut before being hard frozen, although they do not *powder-post*; yet if they are cut in the middle of winter, or in the spring of the year, and the bark is not taken off, the grub will immediately commence its ravages between the bark and the wood. I have walnut timber on hand which has been cut from one to ten years, with the bark on, which was designed for ox-helves and ox-bows, and not a worm is to be found therein; it was cut between 1st August and 1st November. I have other pieces of similar timber cut in the winter months, not two years old, and they are entirely destroyed, being full of *powder-post* and grub-worms."

What shall we say when doctors disagree? The theory given to account for what is assumed to be a fact, is, that as the moon grows the sap rises, and the wood, therefore, is less dense than when the moon is waning, because at that time the sap in the tree diminishes. No evidence whatever can be offered in support of the theory, and one would certainly imagine that the rise or fall of the sap would depend on the quantity of heat which reaches the foot of the tree, and not at all on attraction.

All investigations tend to prove that the only proper time for felling timber is that at which the tree contains the least sap. There are two seasons in each year when the vessels are filled. One is in spring, when the fluid is in motion to supply nutriment to the leaves, and deposit material for new wood; the other is in the early part of autumn, when, after the stagnation which gives the new wood time to dry and harden, it again flows to make the vegetable deposits in the vessels of the wood. At neither of these times should trees be felled; for, if the pores be full of vegetable juices, which being acted upon by heat and moisture may ferment, the wood will decay. Of the two periods, the spring must be the worst, because the wood then contains the greatest quantity of matter in a state fit for germination.

The results of a series of experiments made in Germany show that December-cut wood allows no water to pass through it longitudinally; January-cut wood passed in forty-eight hours a few drops; February-cut wood let two quarts of water through its interstitial spaces in forty-eight hours; March-cut wood permitted the same to filter through in two and a half hours. Hence the reasons why barrels made from wood cut in March or April are so leaky, as the sap is then rising, and the trees are preparing to put forth their leaves.

It thus happens that the time for felling is midsummer or midwinter. The best time for felling, according to some, is midsummer, when the leaves are fully expanded, and the sap has ceased to flow, and the extraneous vegetable matter intended for the leaves has been dislodged from the trunk of the tree by the common sap, leaving it in a quiescent state, and free from that germinative principle which is readily excited by heat and moisture, and if the timber were cut while it remained, would subject it to rapid decay and to operations of worms. Midwinter, amongst some, is chosen as a time for felling, as it is stated that winter-felled heart-wood is less affected by moisture, and likely to be the best and most durable; but as the only peculiar recommendation which that time possesses is the facility which it affords for gradual seasoning, by which timber is rendered less liable to split and get distorted, and slow drying being generally available at any season under shade and shelter, midsummer appears for many obvious reasons the most expedient. In general, all the soft woods, such as elm, lime, poplar, willow, should be felled during winter. In some kinds of trees a little after midsummer appears to be decidedly the best time for felling. Alder felled at that time is found to be much more durable; and Ellis observes, that beech when cut in the middle of summer is bitter, and less liable to be worm-eaten, particularly if a gash be cut to let out the sap some time before felling. Mr. Knowles states that, "About Naples, and in other parts of Italy, oaks have been felled in summer, and are said to have been very durable." Most of the trees in southern Italy are felled in July and August, and the pines in the German forests are cut down mostly in summer time, and it is stated that their wood is sound.

The following are advocates for winter felling, viz. Cato, Pliny, Vitruvius, Alberti, Hesiod, De Saussure, Evelyn, Darwin, and Buonaparte. Some of them consider that winter-felled timber, which has been barked and notched in the previous spring, loses much of that half-prepared woody matter, containing seeds of fungi, &c., that there is no doubt of its superiority to summer-felled timber.

The age at which trees should be felled, and the most suitable time for the work having been determined, there are two other things which claim attention.

The *first* of these is the removal of the bark from the trunk and principal branches of the tree. For, in oak trees, the bark is too valuable to be lost; and as the best period for the timber is the worst for the bark, an ingenious method has been long partially practised, which not only secures the bark at the best season, but also materially improves the timber. This method consists in taking the bark off the standing tree early in the spring, and not felling it till after the new foliage has put forth and died. This practice has been considered of inestimable value; for by it the sap-wood is rendered as strong and durable as the heart-wood; and in some particular instances experiments have shown it to be four times as strong as other wood in all respects similar, and grown on the same soil, but felled with the bark on and dried in sheds. Buffon, Du Hamel, and, in fact, most naturalists, have earnestly recommended the practice. Evelyn states, "To make excellent boards, bark your trees in a fit season, and let them stand naked a full year before felling."

In regard to the time that should elapse between the removal of the bark and the felling of a tree, a variety of opinions exist. It was the usual custom of early architects to remove the bark in the spring, and fell the trees during the succeeding winter. Later investigations seemed to have proved that it is better to perform the work three or even four years in advance, instead of one, although Tredgold appears to think one year too long. Trees will, in most situations, continue to expand and leaf out for several seasons after the bark has been removed. The sap remaining in the wood gradually becomes hardened into woody substance, thereby closing the sap vessels and making it more solid. As bark separates freely from the wood in spring, while the sap is in motion, it should be taken off at that period. When the above method is not adopted, it is well either to pierce the trunk some time before felling to drain out the sap, or immediately on its being felled to set it on end.

The *second* suggestion is, to cut into and around the entire trunk of the tree, near the roots, so that the sap may be discharged; for in this manner it will be done more easily than it can be by evaporation after the tree is felled. In addition to this, if it be permitted to run out at the incision, a large portion of the new and fermentable matter will pass out with it, which would remain in the wood if only such material is removed as would pass off by evaporation. This cutting should be made in the winter previous to the August in which the tree is to be felled; and the incision should be made as deep into the heart-wood as possible without inducing a premature fall of the tree.

The custom of ringing or girdling the tree before felling has been advocated, on the ground that the seasoning is thereby expedited, and also more thoroughly effected. This is doubtful, at least, in oil-containing trees (as teak,

&c.), but the practice appears to be contra-indicated for other reasons: when a tree has been ringed, many woodcutters object to cut it down on account of its increased hardness. This objection might be waived, were it not for another and more serious one which has been adduced. It is believed to be a fact by some that trees felled after girdling have the heart shake increased. It is difficult to explain this, if it be actually the case.

Many suggestions might be made as regards the mechanical operation of felling trees, with which ancient nations were not unfamiliar:

. . . . . "for thou knowest that there is not among us any that can skill to hew timber like unto the Sidonians."—1 Kings v. 6.

But as these operations are familiar to all intelligent workmen, it is only necessary to mention one, viz. the value of removing from the side of the tree such branches as will strike the ground when it falls, and, by wrenching, cleave the grain of the wood, and thereby injure the timber. Such defects, which are often found after the timber has been seasoned, could not be discovered when it left the mill.

In conclusion, we can truly state that the most extensive felling of trees for *one building* only which we have ever heard or read of is the following:

"And Solomon had threescore and ten thousand that bare burdens, and fourscore thousand hewers in the mountains."—1 Kings v. 15.

CHAPTER IV.

ON SEASONING TIMBER BY NATURAL METHODS, VIZ. HOT AND COLD AIR; FRESH AND SALT WATER; VAPOUR; SMOKE; STEAM; BOILING; CHARRING AND SCORCHING, ETC.

ALL timber must, whether it be sap-wood or heart-wood, be placed in situations which will allow the sap to exude or evaporate, and this process is the one technically known by the term "seasoning." There are natural and artificial modes of seasoning, both of which have their recommendations; but the former has certainly the right of preference, as it gives greater toughness, elasticity, and durability, and therefore should always be employed in preparing timber for carpentry. As the word "timber" has been frequently used, it may be as well to state that it is derived, according to Dr. Johnson, from the Saxon, *timbrian*, to build: hence the above definition. The legal definition of timber is restricted to particular species of wood, and custom varies in different countries as to the species ranked among the timber trees.

When a tree is felled, it encloses in its fibres as well as in capillary channels a considerable quantity of sap, which is nothing else but water charged with gummy, saccharine, saline, mucilaginous, and albuminous matters. In this state, the latter are very liable to ferment, but they lose their liability when, by the evaporation of the sap, they pass to a dry and solid state; so that the first suggestion which naturally presents itself to the mind, is to subject the timber to a lengthened seasoning.

But the present demands for time will not admit of this, and therefore it is imperative to resort to artificial and speedy methods.

With respect to the value of timber in the log, owing to its becoming rent by the weather, it sells for 15 per cent. less the second year than the first, and so on for less and less the longer you keep it.

A natural seasoning may be adopted for specimens of moderate thickness, such as deals, planks, &c. At the end of eighteen months from the time of importation they are scarcely dry enough for the consumer's use.

When there is time for drying it gradually, all that is necessary to be done on removing it from the damp ground of the forest, is to place it in a dry yard, sheltered from the sun and wind, and where there is no vegetation; and set it on bearers of iron or brick in such a manner as to admit of a ventilation all round and under it. In this manner it should continue two years, if intended for carpentry; and double that time, if intended for joinery; the loss of weight which should take place to render it fit for the purposes of the former being about one-fifth; and for the latter about one-third. In piling it, the sleepers on which the first pieces are laid should be perfectly level, and "out of the wind," and so firm and solid throughout that they will remain in their original position; for timber, if bent or made to wind before it is seasoned, will generally retain the same form when dried.

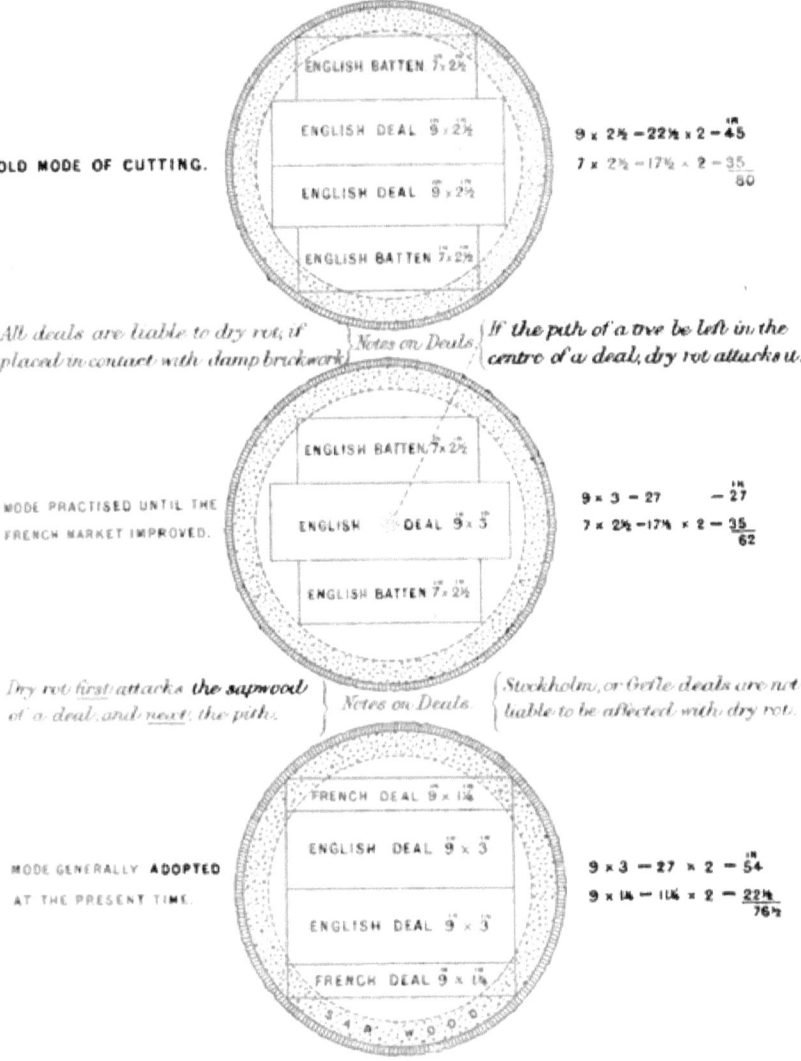

PLANS OF DIFFERENT BALTIC MODES OF CUTTING DEALS FOR THE ENGLISH AND FRENCH MARKETS.
THE SMALLEST TREES ARE CUT FOR DEALS; THE LARGEST FOR LOGS.

OLD MODE OF CUTTING.

All deals are liable to dry rot, if placed in contact with damp brickwork. } Notes on Deals { *If the pith of a tree be left in the centre of a deal, dry rot attacks it.*

MODE PRACTISED UNTIL THE FRENCH MARKET IMPROVED.

Dry rot first attacks the sapwood of a deal, and next the pith. } Notes on Deals { *Stockholm, or Gefle deals are not liable to be affected with dry rot.*

MODE GENERALLY ADOPTED AT THE PRESENT TIME.

"STRONG DEALS rend. When sawn, they do not give saw dust, but the fibres tear.
BEST DEALS are light, mellow, and exhibit a silky texture when planed.
"BESTS". Wholly free from knots, shakes, sapwood, or cross grain, and well seasoned.
"SECONDS". Free from shakes, and sapwood; small knots allowed.
"THIRDS". All that remains after "bests" and "seconds" have been picked out.

Blocks of wood should be put between the "sticks" of timber, and each piece directly over the other, so that air may freely pass through the whole pile; for while it is necessary to shield timber from strong draughts of wind and the direct action of the hot sun, a free circulation of air and moderate warmth are equally essential.

If timber is not used round, it is good to bore out the core; as, by so doing, the drying is advanced, and splitting prevented, with almost no sacrifice of strength. If it is to be squared into logs, it should be done soon after some slow drying, and whole squared, if large enough, as that removes much of the sap-wood, facilitates the drying, and prevents splitting, which is apt to take place when it is in the round form, in consequence of the sap-wood drying before the heart, from being less dense. If it may be quartered, it is well to treat it so after some time, as the seasoning is by that means rendered more equal. It is well also to turn it now and then, as the evaporation is greatest from the upper side. In France, the term "bois du brin," means timber the whole size of the tree, excepting that which is taken off to render it square.

To prevent timber warping to any serious extent, it should be well seasoned before it is cut into scantlings; and the scantlings should be cut some time before they are to be used, in order that the seasoning may be as perfect as possible; and if they can be set upright, so much the better, as then they will dry more rapidly. The white lowland deals of Norway and the white spruce deals of Canada have the same disposition to warp and split on drying.

F

Du Hamel has shown that it is a great advantage to set the timber upright, with the lower end raised a little from the ground; but as this cannot always be done, the timber-yards should be well drained and kept as dry as possible. "Ancient architects," observes Alberti, "not only prevented the access of the scorching rays of the sun and the rude blasts of wind, but also covered the surface with cow-dung, to prevent the too sudden evaporation from the surface." The warping of timber is attributable by some to the manner of its growth. Boards cut out of a tree that is twisted in its growth will not keep from warping; boards cut from trees that are grown in open situations have another fault, in the heart of the tree not running straight like forest-grown wood. In a plank cut from a tree of this kind in a straight line, the heart will traverse it from one end to another. No treatment will prevent it from warping or drying hollow on the side farthest from the heart. Where the heart is in the centre of a plank, and each side has an equal chance of drying, it will not warp; but there will be a shake or crack upon each side, denoting the position of the heart.

Some deals, and particularly the stringy deals, are very hygrometric, and never lose the property, however long they have been seasoned, of expanding and contracting with change of weather. White Petersburgh deals are said to have that property, however long they may have been kept, so that if used in the panel of a door, the wood alternately enters and recedes from the groove into which it fits, as the paint will show when that kind of deal has been used for a panel.

The wood of the north side will not warp so much as the wood from the south side. The face of the planks should be cut in the direction which lay from east to west as the tree stood. If this be done, the planks will warp much less than if cut in the opposite direction. The nature of the tree, the soil upon which it is grown, the position of its growth, the period of the year in which it is felled, and the length of time between its felling and converting, are the principal points to be considered; a thorough knowledge and study of which is the only true principle on which we can hope to deal with the warping and converting of timber.

Wood, when it is cut into small pieces, very soon acquires its utmost degree of dryness. Dr. Watson, Bishop of Llandaff, in the month of March, cut a piece from the middle of a large ash tree that had been felled about six weeks, and weighed it; its weight was 317 grains. In seven days it lost 62 grains, or nearly one-fifth of its weight. It was weighed again in August of the same year, but had not lost any more of its weight; hence it had become perfectly dry in the short space of seven days. He also found that the sap-wood of oak lost more weight in drying than the heart-wood, in the proportion of 10 to 7.

The time that is required to season or dry a piece of timber obviously depends upon its magnitude; as a *general rule*, large timbers will not continue good so long as small ones, as sufficient time is rarely given for a thorough seasoning. The time required to dry a piece of timber, all other things being alike, will depend on the quantity of surface exposed to the action of the air;

therefore, while the quantity of timber remains the same, the larger the surface, the sooner it will dry. Also, if the quantity of surface remains the same, the time of drying will be proportioned to the quantity of matter; as the greater the quantity of matter under the same surface, the longer it will be in drying.

As drying proceeds most rapidly in small pieces, it is therefore important to reduce the timber to its proper scantlings or size for use; for however dry a piece of timber may be, when it is cut to a smaller scantling it will shrink and lose weight, being always less dry in the centre than at the surface; and the more rapidly the drying has been carried on, the greater will be the difference. Nevertheless, in the first stage of seasoning it is best that it should proceed slowly; otherwise, the external pores shrink so close as not to permit the full evaporation of the internal moisture, and the piece would split from unequal shrinking; and lastly, it should be reduced to the proper scantling, as already observed, some time before it is to be framed. Various tables have been given by writers on timber, the result of algebraical calculations, of the times of seasoning and drying for different woods of different lengths, breadths, and thicknesses, in the open air; but as wood even of the same description and quality varies so much, this matter is best left to those who are well acquainted with timber. It may, however, be stated that the time required for drying under cover is shorter than in the open air, in the proportion of 5 to 7.

The English shipwright considers that three years are

required to thoroughly season timber. The timbers for ships are usually cut out to their shape and dimensions for about a year before they are framed together, and they are commonly left a year longer in the skeleton shape to complete the seasoning, as in that condition they are more favourably situate as regards exposure to the air than when they are closely covered in with planking.

It is worthy of mention that all the harder woods require increased care in the seasoning, which is often badly begun by exposure to the sun or hot winds in their native climates: their greater impenetrability to the air the more disposes them to crack, and their comparative scarcity and expense are also powerful arguments on the score of precaution. Oak timber requires to be very carefully seasoned, as it is generally used in buildings for the best description of work, and should unseasoned oak be used for "panelling," any shrinkage will be fatal to the work. Mr. George Marshall, timber merchant (see the *Builder*, January 20, 1872), with respect to seasoning oak timber, observes: "I should select oak trees known to be old and hearty, with clean, straight butts, from 15 inches to 20 inches in diameter. I should then have the bark taken off as they stand, and leave them thus till the winter; the sap will then partially dry out, and make the wood a rich brown colour. As soon as they are cut down, have them sawn up at once into the lengths you require the panelling, 6 inches or 8 inches wide and 1 inch to $1\frac{1}{2}$ inch thick. Be careful to cut all the heart shakes, by having one cut through the centre of the log before edging the boards to the required width. With regard to the drying process, stack the boards in a shed with a good draught through it, and load them down, with slips between each board, to prevent warping. If this be done they will be found to dry well and speedily, and they will not require to be exposed to the weather."

Mr. Robert Phillips, on seasoning oak for panelling, states: "If the tree is large enough for the purpose, cut it into four, in sections, by drawing a vertical and horizontal line across the end, meeting in the centre. If too small for this, cut it into $4\frac{1}{2}$-inch or 6-inch plank, as soon as possible after felling, and then stack *on end* out in the open: do not lay on the ground, but stand it as nearly *vertical* on its end as possible, and keep it *wet* during the first three months. If the weather is dry, well wet it with water poured on the top, and allowed to run down. Let the ends stand on a piece of quartering, to keep it out of the dirt, or it will be stained some distance up. After standing thus for some six months, after putting it in a dry place for some time, cut it into the scantlings you require, always bearing in mind that oak will, after this seasoning, shrink at least half an inch to a foot, in width and thickness. They should then be stacked and stripped, and covered with spare boards, and weighted on the top, for at least six months—as much more as possible—in a covered shed, with plenty of air, occasionally turned over and shifted, till they are dry enough to make dust when planed, and not turn the shaving black. They will then be fit for use.

"I should advise for the panels to be cut *feather-edged* boards, in radial lines from the centre of the tree: it will be a waste of material, but will repay in the beauty of the wood, and the way it will stand without warping. Most of the panels of our old cathedrals were rent (*not sawn*) in this way, and stand admirably. The butt of the tree should be taken, the top being used for a rougher purpose."

Mr. George Marshall and Mr. Robert Phillips might have mentioned that the oak trees should be of the *Quercus Robur* species, and *not* the *Quercus Sessiliflora*. They are easily distinguished when growing by the following peculiarities: The acorn-stalks of the *Robur* are **long**; the acorns grow singly, or seldom two on the same footstalk; the leaves are **short**. The acorn-stalks of the *Sessiliflora* are **short**; the acorns grow in clusters of two or three, close to the stem of the branch; the leaves are **long**.

WATER SEASONING.

When there is not time for gradual drying, the best method, perhaps, that can be adopted, especially for sappy timber, and if strength is not principally required, is immediately on felling to immerse it in running water; and after allowing it to remain there about a fortnight, to set it in the wind to dry. Some persons prefer this method of seasoning timber, as they say it prevents cleaving, and strips and seasons better afterwards. This process has been adopted with good results by placing the boards end on at the head of a mill race for fourteen or twenty days, at most, and then setting the boards upright, and subject to the action of the sun and wind; though it is questionable whether the sun will not do them more harm than good. As they stand, turn them daily, and when perfectly dry—

which process will take about one month—it is considered they will be found to floor better than timber after many years of dry seasoning. The sap-wood of oak is said to be improved by this method, being much less subject to be worm-eaten; and providing it is placed in fresh running water, Mr. G. A. Rogers, the celebrated wood carver, is of opinion that the colour of the oak is improved. The more tender woods, such as alder and the like, are less subject to the worm when water seasoned. Beech is said to be much benefited by immersion. It should be remembered that the timber should be altogether under water (chained down beneath its surface), as partial immersion is very destructive. Du Hamel considers "that where strength is required, wood ought not to be put in water." Timber should never be kept floating in ponds or docks, as in London; but it should be stacked, as at Liverpool and Gloucester. Timber that has been lying for months in ponds or docks is sometimes cut up, and in six or seven days fixed in a building; consequently, the usual result takes place, viz. dry rot. After having been swelled by soaking much beyond its former bulk, the baulk of timber is put on the saw-pit, and cut into scantlings, and framed while in this wet state, therefore it cannot be surprising that the dry rot soon appears as a natural consequence.

Amongst wheelwrights the water seasoning is in general favour. It is said that the colour of the white woods is improved by water seasoning, boiling, or steaming. The Venetians place the oak used for gun-carriages in water for two years before it is used, and the timber for sea service two or three years under water. The Turks do not appear to pay any attention to seasoning, for they fell their timber at all times of the year without any regard to the season, and although they grow very good oak, it is used so green and unseasoned that it not only twists, but decays rapidly, as anyone may observe in the houses at Constantinople and other Turkish towns.

Timber is rendered more durable by placing it in a stream of water, saturated with lime, for eight or ten days, and it also makes it less liable to the attack of worms; but it, however, becomes hard after being dried, and is difficult to be worked; and therefore the process should be applied to timber which has been sawn into scantlings, and is ready for use. Mr. William Chapman, in 1812, considered that an immersion of timber in hot limewash in deep ponds, exposing little surface to the air, merited a trial; but in 1816, from experiments he had made, he was of opinion that it had proved injurious to timber.

Evelyn states that green elm, if plunged for a few days in water (especially salt water), obtains an admirable seasoning. According to Society of Arts *Trans.*, 1819, every trace of fungus was eradicated from the ship "Eden," by its remaining eighteen months under sea water. Salt water is considered good for ship timber, but for timber to be employed in the construction of dwelling-houses, fresh water is better. Pliny notices, as a fact, that certain woods on being dried after immersion in the sea acquire additional density and durability. M. de Lapparent, late Director of the French Navy, considers that timber cannot be seasoned in salt water, but in fresh, or at the most, in brackish water. The condition of the timber which, at the port of Rochefort, is kept in ditches filled with fresh water is in this respect most favourable; that kept at Toulon, Brest, and Lorient, where the water is brackish, is much less so; but to estimate their relative advantages, it would be necessary to test the average density of these waters. It is, however, at Cherbourg that this natural preparation of timber is the most inefficient, as the beds of sand in which the timber is buried, near the Pool of Tourlaville, contain but a small quantity of water, which, being nearly always stagnant, very quickly exhausts itself, and is very prejudicial.

At the Cologne International Agricultural Exhibition, in 1865, three sleepers were exhibited from the Magdeburg Leipsiger Railway, from the Salt Work Branch, at Stassfurt, laid in 1857. These were moistened by the refuse of the salt which was lost from the load and by the rain. The jury in their Report stated that these sleepers proved nothing, "because every old table on which fish or meat has been salted, proves that a constant moistening with salt water preserves the wood from decay, but as soon as the process of salting is given up, the salted matter is immediately given out, and the timber soon decays. In this case it would have been important to have known that these sleepers, after having been salted, had lain anywhere else than in the Salt Work Branch without getting fresh salt applied, and then to have seen if they would have been as perfect as they are now. They, indeed, prove nothing but the fact that if sleepers be daily sprinkled with salt they will remain sound, but the price paid for this durability might be very considerable." As the use of salt as a preservative agent will be considered in the next chapter, it will be best to defer the consideration of salt-water seasoning until then.

In India, teak, sal, and blackwood, &c., improve by lying in water, or in the soft black mud of an estuary: there is one exception, viz. heddé, which deteriorates from steeping, and should be carted to its destination.

Evelyn states that he had found a fortnight's immersion in river water sufficient, and this opinion is held by Silloway, a North American authority; but Dr. Porcher, a South American writer, recommends a six months' immersion in water, and a six months' exposure to wind and shade. Vitruvius and Alberti consider that timber should be left immersed in a running stream thirty days.

It is considered that the longer wood has remained under water, the more rapidly it dries; for instance, every one is aware that the firewood brought out of the river is less green and burns better than that brought by waggon or boat.

In 1817, Admiral Count Chateauvieux, a Sardinian naval officer, observed to Mr. McWilliam that it was a custom at the Royal Arsenal, at Genoa, as a preventive against the diseases of timber, to steep it about three years in fresh water immediately after it is felled. Mr. James Dickson, of Gottenburg, timber merchant (member of the firm Peter Dickson and Company, London), many years engaged in the Swedish timber trade, observed in 1835, "If square timber lies in the water two or three years, it rends at the heart, but I should not say it would, perhaps, for the first year; but the exterior part rends soon by exposure to the weather." In 1818, the Chevalier de Campugano, Secretary of Legation to the Spanish Embassy, stated that in Spain, when timber is felled, it is generally laid in water for a considerable time.

The sap in timber, by reason of the matters which it holds in solution, is denser than pure water; moreover, it is enclosed in fibres or channels permeable at the ends.

Supposing in submerged timber, the surrounding water to be flowing, or at least changing, this water will conclude by occupying, if not altogether, at least in a great degree, the place of the sap, which will have issued forth, carrying with it the fermenting principle with which it is charged. The timber, therefore, which has remained *sufficiently long* in the water ought to be much less susceptible of fermentation than that seasoned only by the atmosphere. Besides, as pure water evaporates much easier than that which contains certain principles, this timber ought to be seasoned much sooner than the other.

Of steeping generally, whether in cold or warm water, it must be particularly observed that it dissolves the substance of the wood, and necessarily renders it lighter; indeed, it is known that notwithstanding wood that is carefully submerged remains good for a very long period after the water has dissolved a certain soluble part, it is, when taken out and dried, liable to be brittle, and unfit for any other work but joinery.

SEASONING BY STEAMING AND BOILING, ETC.

For the purposes of joinery, steaming and boiling are very good methods, as the loss of elasticity and strength which they produce, and which are essential in carpentry, is compensated by the tendency to shrinkage being reduced; the durability also is said by some to be rather improved than otherwise, at least from steaming. If steaming be not carried on too quickly it will answer, but if it be pushed with too much vigour it is very apt to produce a permanent warping and distortion of the material. Oak of British growth may be seasoned by this process, as without this precaution it requires a long time to season. It has been ascertained, that of woods seasoned by these methods, those dried soonest that had been steamed; but the drying in either case should be somewhat gradual, and four hours are generally sufficient for the boiling or steaming process. The question of time will depend upon circumstances: some persons consider that one hour should be allowed for every inch in thickness. In some dockyards, salt water is used in the boilers, in others fresh, from considerations of convenience; and the fact is, plank boiled in salt water never gets rid of the salts that naturally enter the pores of the wood in boiling; and such being the case, the ship in which this plank is used is much more liable to the effects of damp than she would have been if the plank had been boiled in fresh water.

Boiling and steaming are likewise employed for softening woods, to facilitate the cutting as well as bending of them. Thus, in Taylor's patent machines for making casks, the blocks intended for the staves are cut out of white Canada oak to the size of 30 inches by 5 inches and smaller. They are well steamed, and then sliced into pieces $\frac{1}{2}$ inch or $\frac{5}{8}$ inch thick, at the rate of 200 in each minute, by a process far more rapid and economical than sawing; the instrument being a revolving iron plate of 12 feet diameter, with two radial knives arranged somewhat like the irons of an ordinary plane or spokeshave.

How far steaming or boiling affects the durability of timber has not been satisfactorily ascertained; but it is said that the planks of a ship near the bows, which are bent by steaming, have never been observed to be affected with dry rot. With respect to boiling, Du Hamel's opinion is not favourable as to its adding to the durability of timber; for when a piece of dry wood was immersed in boiling water, and afterwards dried in a stove, it not only lost the water it had imbibed, but also a part of its substance; and when the experiment was repeated with the same piece of wood, it lost more of its substance the second time than it did the first. Tredgold—no mean authority —considers that "boiled or steamed timber shrinks less, and stands better than that which is naturally seasoned." Barlow is of opinion that "the seasoning goes on more rapidly after the piece is steamed than when boiled."

At the close of the Crimean and Baltic campaigns the port of Cherbourg was almost completely cleared of staves sufficiently seasoned for making casks. The engineer at the head of the coopering department determined to boil in fresh water the newly-cut staves, and compare the time of their seasoning with that of other staves cut from the same forests, but not prepared; and the result was that after four or five months' exposure to the atmosphere, the boiled staves were perfectly fit for working up, while to bring the others to the same point fifteen months were barely sufficient.

Steaming is understood to prevent dry rot. No doubt boiling and steaming partly remove the ferment spores, but *may not* destroy the vitality of those remaining. For, according to Milne-Edwards, on 'Spontaneous Generation,' he has seen tardigrades resist the prolonged action of a temperature of 248° Fahr., and has known them to survive a temperature of 284° Fahr. That low forms of vegetation are fully as tenacious of life cannot be doubted.

Boiling and steaming also coagulate the albumen at 140° Fahr. Although coagulated albumen is insoluble in water, the water solution is by this heating process sealed up in the wood, and the cohesion of the latter is said to be diminished.

The first essays in the art of drying wood artificially carry us back to a period now tolerably remote. Wollaston and Fourcroy both recommended the drying of wood in ovens. Newmann, a German chemist, suggested another method, which has since been adopted in a somewhat different form, i. e. *steaming* the wood. Newmann placed the wood to be dried in a large wooden chest, taking care to leave spaces between the pieces, and then turned on the steam from a boiler provided for the purpose. The condensed steam, charged with albuminous matter taken up from the wood, or rather from its surface, was run off from time to time, and the process of the operation was judged by the colour of the water. When the latter was clear and colourless the chest was opened, and the wood withdrawn for use without further preparation. The process would have been useful enough if *superheated* steam, which would have dried the wood by absorbing the moisture, could have been used, but the cost of the process would doubtless have been too high to permit of its practical application.

In 1837, M. de Mecquenem devised a method of desiccation, in which the pieces of wood to be dried were placed in a closed chamber, and subjected to a current of hot air, heated for the purpose by a special apparatus, and driven by a blower. The air entered by apertures in the lower part of the chambers, and escaped at the top laden with the moisture absorbed from the wood.

In 1839, M. Charpentier obtained a *brévet d'invention* for a process of drying wood in hermetically-closed chambers. The wood was subjected to the action of air heated by contact with metal plates covering the flue of a coke furnace. This air entered by conduits on the level of the floor of the chamber, and escaped at the top through apertures leading into the chimney of the furnace.

In the same year, M. Saint Preuve invented a process for forcing steam into pores of the wood, and, by condensation of this steam in the pores, sucking in a preservative preparation.

In 1847, MM. Brochard and Watteau's process was introduced. It consists simply of filling the cylinder with steam, and making a vacuum by forcing in a cold solution of salt, &c.

The plan which has been for some years in use in

England is the injection, by means of a ventilator, of hot air into the drying stove where the wood is placed: by this the temperature is gently and gradually raised until it reaches boiling heat. But, as wood is one of the worst conductors known of caloric, if this plan is applied to large logs, the interior fibres still retain their original bulk, while those near the surface have a tendency to shrink; the consequence of which would be cracks and splits of more or less depth.

Timber may be dried by passing rapid currents of heated air through it under pressure. This plan was carried out with the timber used for the floorings of the Coal Exchange, London. The wood was taken in its natural state, and in less than ten days it was thoroughly seasoned. In some cases, from 10 to 48 per cent. of moisture was taken out of the wood, and although the floorings have now been down a great many years, it is stated that very little shrinkage has been found, except in the case of a few pieces which were put down in the latter portion of the work, and which had not been submitted to the seasoning process.

The process of desiccation, patented by Messrs. Davison and Symington, in 1844, is of great practical value in reducing the time requisite for seasoning timber. It is peculiarly applicable to the seasoning of flooring boards and of the wood used in joiners' work. Care must be exercised when removing the timber from the stove to the building in which it is to be used, that it be not exposed to the wet, nor even to a damp atmosphere for any lengthened period. The advantage of this process over the ordinary stoving consists in the temperature never being so high as to scorch the wood, by which the strength of the fibres would be injured; and in the facility for removing the vapour as fast as it is expelled from the wood, in consequence of the air being propelled through the stove at any required velocity and temperature. As compared with furnace and steam-stoving ordinarily employed to desiccate woods, the great superiority of this process is established by its seasoning the wood quite as rapidly, but much more thoroughly; and instead of wood being rendered brittle, as it is to some extent by stoving, this mode does not reduce the strength and tenacity of the wood. The principle of the invention is propelled currents of heated air; but the heat has to be regulated according to the texture of the various woods. Honduras mahogany might be exposed to a heat of 300°, and the whole of the moisture can be taken out in three days. Timber 9 inches square is considered by Mr. Davison a proper size for his invention. This process is described as "A method or methods of drying, seasoning, and hardening wood, and other articles, parts of which are also applicable to the desiccation of vegetable substances generally." The first or principal part of the invention consists in drying, seasoning, and hardening wood and other articles—among which other articles are included generally all things made of wood, or chiefly of wood—by means, as has been stated, of rapid currents of heated air. The manner in which these currents of heated air are produced, is by an apparatus consisting of a furnace and a series of pipes withinside of a core of brickwork.

On each side of the furnace, on a level with the fire-bars, is a horizontal tube; communicating and springing from these tubes are a series of eighteen tubes placed vertically and parallel to each other over the furnace. The outer end of one of the horizontal tubes communicates with a fan or other impelling apparatus for driving a constant stream of atmospheric air through the tubes. As the air passes through the tubes it becomes heated at a high temperature, and rushes out at the farther end of the other horizontal tube, and is thus conveyed to the place where it is applied. The materials to be subjected to the heated currents, such as logs, deals, &c., by outward application, must be placed in closed chambers, galleries, vaults, or flues, which are to be of any suitable form or magnitude; but it is recommended that they should be made of fire-brick, and have double doors or shutters for introducing or removing the wood. Honourable Mention was made of Messrs. Davison and Symington's process of desiccation, by the jury, Class IV., Exhibition of 1851, England.

Some amusing instances are related of the efficiency of Davison and Symington's process. Thus, a violin had been in the owner's possession for upwards of sixteen years; how old it was when he first had it is not known. Upon being exposed to this process it lost, in eight hours, no less than five-sixths (nearly five and two-thirds) per cent. of its own weight. This there is every reason to believe was owing to the blocks glued inside, for the purpose of holding the more slender parts together. A violin maker of high reputation, having an order to make an instrument for one of the first violinists of the day, was requested to have the wood seasoned by this process; only three days were allowed for the experiment, in which the wood was seasoned and sent home. The two heaviest pieces were reduced in weight $2\frac{1}{2}$ lbs. It is ascertained that, by this means of drying, the effect of age has been given to the instrument made from the above wood, and it was, in 1848, *first fiddle* in the orchestra of Her Majesty's Theatre, London. The wood had been in the possession of its owners for eight years, and it was sent from Switzerland, in the first instance, as *dry wood*.

In proof of the value of this invention for the manufacture and cleansing of brewers' casks, it was stated, in 1848, that since its adoption at Trueman's brewery, Spitalfields, a saving of 300 tons of coals has been effected annually.

Flues or chambers for the heated air may be constructed in parallel lines, either in the floors or upright walls of a building, having narrow openings through which the heated air may issue in thin streams, and spread itself over the surface of the wood. If the openings are in the floor, the wood will require to be placed in an upright position; but if admitted in a horizontal direction, standards and skeleton shelves will be necessary to lay it upon. The great object, in all cases, is to bring the heated air as speedily as possible into contact with the wood, and to allow it, after it has done its office, to pass away as speedily.

Furnaces and apparatus for the production of rapid currents of heated air may be erected to prepare any quantity of timber or articles of wood at one time, but care should be taken that whatever the size of the outlet may be from the series of pipes or vessels by which the heat is generated, an outlet of at least equal dimensions is left for the free exit of the air and the vapours thrown off. It should also be observed, in constructing the open space in the floor or upright walls for the stream of heated air to pass towards the timber, that the superficial area of the whole of them combined does not exceed the dimensions of the principal outlet of the pipes at the extremity of the furnace, so that a free current of heated air may be allowed to pass uniformly throughout the chambers containing the wood to be prepared. The temperature proper to be given to the air, and velocity to the current in each case, will depend on the size, density, and maturity of the wood to be acted upon. The inventors found by their experiments that wood generally may be advantageously subjected to currents of air raised to a temperature of 400° Fahr., when the currents are impelled at the rate of 100 feet per second. But when the wood is in a green state, it is better to commence at a lower temperature, say from 150° to 200°, and gradually raise it to the high degree before stated, as the desiccation proceeds, an object which may, in some cases, be facilitated by carrying a cold-air drain from the fanner or other propelling apparatus, and attaching a damper to it, so that any quantity of cold air required to reduce the temperature of the hot current may, from time to time, be admitted. When, again, the wood is in the log or unconverted state, it should be bored or augured out in the centre, and the current of hot air caused to traverse it as well interiorly as exteriorly, whereby much time will be saved in the process of desiccation, and a more uniform result obtained.

Woods treated in this manner, and with the above modifications when requisite, part rapidly with their natural sap and any other aqueous matter which they may contain, and the fibres are brought closer together.

With respect to the time required to season the wood upon this plan, much must depend upon the original state of dryness it may be in, as well as the quality and temperature of the heated air forced into contact with it. It may suffice to remark that the wood may safely remain thus exposed till any escape of moisture ceases to be perceptible. This may be readily known, either by applying a mirror or any polished surface to the outlet, or by calculating the quantity of moisture removed from the wood, which will be found to range between $\frac{1}{4}$ and $\frac{1}{15}$th of its whole weight. For the purpose of ascertaining more correctly the amount of moisture removed from time to time, when the wood is placed in seasoning chambers as already described, an opening should be constructed in the chamber, in any convenient position, through which a specimen of the wood may be withdrawn and weighed.

Between 1848 and 1853, Mr. Bethell, who had paid much attention to the subject, obtained several patents, both in England and France, for stoves for drying wood. In his English patent of 1848, and the subsequent French one of 1853, we find a description of a peculiar kind of stove, on the following plan:

It consisted of a rectangular chamber formed of three walls and vaulted over, the whole in brickwork, with a certain thickness of slag in the centre, to prevent loss of heat. One extremity of the chamber was open to admit of the introduction of the wood by means of a truck running upon longitudinal iron rails. The opening was closed with a double door when the chamber was full. On the exterior of the opposite end of the chamber was a furnace to burn coal, coke, wood, or tar, according as it was desired to *dry* the wood simply, or, in the words of the inventor, to *smoke* it, i. e., to impregnate it with the antiseptic gaseous matters evolved in the imperfect combustion of certain tarry substances. The heated air or smoke entered through a flue running along the floor and branching at the end, and it escaped, or was pumped out, at the top of the vaults. Bethell considered that the interior of the chamber should be kept at a temperature of 110° Fahr., and that the duration of the process should be regulated by the condition of the wood. His experiments showed that this time varied from eight to twelve hours, the rapidity being attained at the cost of a relatively large expenditure of fuel. In point of fact, the draught was too great to permit of the utilization of the full amount of heat contained in the gaseous matter, which escaped at a temperature very little below that at which it entered. The heat produced by the fuel was badly utilized, and it is open to question whether, under any circumstances, large pieces of wood, such as sleepers, could be dried in so short a time as eight or twelve hours. The drying could only be effected by the use of a very high degree of temperature, tending to split the wood and weaken its strength. This view was confirmed by the results obtained in a long series of experiments made, in 1852-3, by an English manufacturing company, known as the Desiccating Company. A low temperature, and long continuance of the drying process, appear to be the conditions essential to the success of artificial desiccation, particularly with wood intended for cabinet-making, turning, joinery, ornamental work, &c., in which it is desirable, as far as possible, to prevent splitting, warping, and other changes of structure in the material. These results, it would seem, were not secured by the arrangements above described.

Some years since, a stove was constructed for Messrs. S. and J. Holme, very extensive builders at Liverpool, for the purpose of drying timber for floors, and other fittings of houses, &c., by the application of Messrs. Price and Manby's patent warming apparatus; the want of seasoned timber, with the great number of men they employed, being a serious inconvenience and loss. In their large undertakings Messrs. Holme found a difficulty in keeping a stock of dry timber. The dimensions of the stove in which the timber was to be dried was 43 feet long, 11 feet wide, and 17 feet 6 inches high, and the cost of the apparatus was about 150*l*. It was calculated to hold about 30,000 superficial feet of 1-inch boards, which, upon the steam-pipe system, occupied full three weeks in drying. This apparatus of Messrs. Price and Manby, with rather less fuel, was considered to thoroughly dry each stove-full in ten days, thus saving a consumption of ten days' fuel, independent of the advantages of expediting business. The average temperature was 104°, and as the continuous stream of pure air passing between the metallic plates was divested of its moisture, it carried off the dampness of the timber in an imperceptible manner. An experiment was tried, by having a flooring batten, 7 inches by $1\frac{1}{4}$ inch, cut from a piece of timber which had been floated, and was as full of water as it could be, placed in the stove; and when the temperature was 102°, it remained there five days, and when sawn down into $\frac{5}{8}$ inch thick, and planed, it was found to be perfectly dry throughout. The heat was so gentle, and the evaporation so equal, that the timber was never rent, as when exposed to the air and a hot sun: in short, Messrs. Holme considered it one of the most perfect timber stoves that had been made.

It may be remarked with respect to desiccation, that the timber to be artificially dried is generally exposed to a great heat for a short time, rather than to a moderate heat for a lengthened one; and the air, saturated with the vapour thus produced, is generally very imperfectly removed. Wood so treated is almost sure to split, from the unequal contraction to which it is exposed; and the pores are also very liable to reopen on the wood being withdrawn from the stove, because there is no gradual and permanent change in their mechanical structure. It is only within the last few years past that the artificial desiccation of wood, before its impregnation with an antiseptic preparation in closed vessels, has been frequently adopted in practice.

We cannot give a better termination to the few remarks we have made about "steaming and boiling timber," than by quoting the opinion of the late Sir Charles Barry, R.A., architect to the new Houses of Parliament, which we propose doing in the following manner:

"York Road, Lambeth,
Nov. 30, 1844.

Sir,

'In reply to your application, we beg to acquaint you that we are willing to undertake the ordinary works required in the finishings of the new Palace of Westminster .... The wainscot to be used in the joiner's work is assumed to be from the best Crown Riga wainscot in the logs, and from pipe-staves of the best quality, in equal proportions, to be *prepared for use by steaming*, or otherwise ....

Grissell & Peto."

Charles Barry, Esq.

Sir Charles Barry recommended this tender to the Treasury for acceptance; but we fancy that he was doubtful about the efficacy of steaming, as we think will appear from the following extract from an "Agreement between Sir Charles Barry and Messrs. Grissell and Peto, builders:

"*First.* That the wainscot is assumed to be from the log and pipe-staves in equal quantities; the prime cost of which, in inch boards, *seasoned by steam*, or other artificial means, so as to be fit for use, is calculated at $6\frac{1}{2}d.$ per foot superficial.

"*Secondly.* That *if it should be found necessary to make use of thoroughly dry wainscot boards* for the whole or any portion of the joiner's work, seasoned by natural means (viz. exposure to the atmosphere), the prime cost of such boards, with the addition of a profit of 7½ per cent., is to be allowed for them, over and above the price of 6½d. per foot superficial, the prime cost of wainscot boards provided for in the contract, as above stated." (Italics are our own.)

SEASONING BY SMOKE DRYING.

Smoke drying in an open chamber, or the burning of furze, fern, shavings, or straw under the wood, is said to give it hardness and durability; and, by rendering it bitter, destroys and prevents worms. It also destroys the germ of any fungus which may have commenced. It is an old and well-founded observation that smoke drying contributes much to the hardness and durability of woods. Virgil appears to have been aware of its utility, when he wrote the passage which is thus translated by Dryden:

> "Of beech, the plough-tail, and the bending yoke,
> Or softer linden, hardened in the smoke."—GEORGICS, i., 225.

Beckman, in his 'History of Inventions,' quotes a passage from Hesiod to the same effect; and adds, "as the houses of the ancients were so smoky, it may be easily comprehended how, by means of smoke, they could dry and harden pieces of timber." In this manner were prepared the pieces of wood destined for ploughs, waggons, and the rudders of vessels:

> "These long suspend, where smoke their strength explores,
> And seasons into use, and binds their pores."—VIRGIL.

The late Brigadier-General Sir Samuel Bentham bestowed much time and attention in endeavouring to ascertain the quickest and best means of drying oak. In his letter to the Navy Board, 6th March, 1812, he says: "By exposing block shells to the smoke of burning wood, they become in the course of two or three days well seasoned in every respect, hard, bright coloured, and, as it were, polished. But it was found in a very short time that the acid with which the shells were thus impregnated very rapidly corroded the iron pins which passed through them.

"In Russia many small articles, such as parts of wheels, wheel carriages, and sledges, are prepared in this manner; so are wheels, at least in some parts of America; and sabots and other small articles in France."

In speaking of artificial heat, he says, in the same letter:

"From all the opportunities I have had of examining the state of timber so prepared by artificial heat, the due seasoning without cracking has appeared to depend on the ventilation happening to be *constant*, but very *slow*, joined to such a due regulation of the heat as that the *interior* of the timber should dry, and keep pace in its contraction with the outer circles."

Mr. T. W. Silloway, in 'American Carpentry,' remarks: "If timber be dried by heat, the outside will become hardened, and the pores closed, so that moisture, instead of passing out, will be retained within."

Bowden remarks, "that the timbers of a small ship underwent the process of charring, either by suspending them over a fire of chips, or by burning the exterior with red-hot irons, so as to char the external surface. Air trunks were also formed between the timbers, for the purpose of evaporating moisture. The state of this vessel was examined five years after she was launched, and it appeared that, although the timbers had been very strongly charred, fungi had grown to a considerable extent on both sides abaft the fore channels, and that the plank near the magazine was completely decayed." *The power of vegetation broke through the incrusted barrier against external affection.*

A method is in operation at Tourlaville, near Cherbourg, for which the inventor, M. Guibert, has taken out a patent, and it is said to give at once more expeditious and sure results than those obtained from the use of dry and hot air. It consists in filling the drying-stove with smoke, produced by the distillation of certain combustible matters, such as saw-dust, waste tan, and smiths' coals, &c. By means of a ventilator, ingeniously arranged, a rotatory movement round the logs laid to season is given to the smoke, so as to obtain an average uniform temperature in every part. By this plan, as the distillation of combustibles is always attended with a considerable discharge of steam, all cracks and splits are said to be prevented.

There is much force in Sir Samuel Bentham's observations respecting the drying of timber by artificial heat: it is certainly not well to attempt to dry it *too* quickly, for if it be subjected to great heat, a large portion of the carbon will pass off, and thereby weaken the timber. Timber too suddenly dried cracks badly, and is thus materially injured: planks of larch or beech are liable to warp and twist if their drying is hastened.

STOVE DRYING.

In some of the large manufactories for cabinet work, the premises are heated by steam pipes, in which case they have a close stove in every workshop heated many degrees beyond the general temperature, for giving the final seasoning to the wood; for heating the cauls; and for warming the glue, which is then done by opening a small steam pipe into the outer vessel of the glue-pot. The arrangement is extremely clean, safe from fire, and the degree of heat is very much under control.

In some manufactories, the wood is placed for a few days before it is worked up in a drying-room heated by means of stoves, steam or hot water, to several degrees beyond the temperature to which the finished work is likely to be subjected. Such rooms are frequently made as air-tight as possible, which appears to be a mistake,* as the wood is then surrounded by a warm but stagnant atmosphere, which retains whatever moisture it may have evaporated from the wood.

Fire-stoves for drying the timber were placed in the magazine, bread-room, and other parts of the 'Royal Charlotte' ship; and the evil of this practice was soon shown, for the vessel became dry rotten in *twelve months*.†

Wood sometimes undergoes a baking process for veneering. Fourcroy has recommended baking timber in an oven, and he has asserted that it would render timber more durable; "but," says Boyden, "it should be subjected to a very strong heat, lest in endeavouring

---

* See white faces of workmen.
† See London newspapers, July, 1812.

to prevent vegetation, we should give it birth." Captain Shaw* observes: "Any artificial heating which burns the air is most injurious to wood and all combustible materials, and renders them much more inflammable than they would be if only exposed to the temperature of the atmosphere."

SEASONING BY SCORCHING AND CHARRING.

Scorching and charring are good for preventing and destroying infection in timber, but have to be done slowly, and only to timber that is already thoroughly seasoned; otherwise, by incrusting the surface, the evaporation of any internal moisture is intercepted, and decay in the heart soon ensues; if done hastily, cracks are also caused on the surface, and which, receiving from the wood a moisture for which there is not a sufficient means of evaporation, renders it soon liable to decay. Charring has little or no control over internal corruption, though it is a good preventive against external infection: it increases the durability of dry, but promotes the decay of wet timber. Farmers very often resort to this method for the preservation of their fence-posts; the charring should extend a little above their contact with the ground. Unless they discriminate between green and unseasoned timber, these operations will prove injurious instead of beneficial.

We have already quoted Sir Charles Barry in favour of *steaming* wood; we now intend giving the opinion of a former pupil of his with regard to *charring* it. Mr. George

---
* 'Fire Surveys,' p. 58.

Vulliamy, architect to the Metropolitan Board of Works, in a specification for oak fencing which was fixed round the boundaries of Finsbury Park, London, in 1867, writes as follows:—" Dig out the ground for the upright standards where shall be directed, and fill in and ram round same with dry burnt earth, stones, and rubbish (the burnt clay will be provided); enclose the boundaries of Park, as shall be directed, with *dry and well-seasoned* heart of English oak, *wrought* upright standards, 6 inches by 5 inches, and 8 feet 6 inches total length, with cut and *splayed* tops, holes drilled for oak pins, and mortised for horizontal rails, as shown on detailed drawings; to stand 5 feet 3 inches out of ground, and *the ends in ground to be well charred before fixing*." (The italics are our own.)

Our ancestors used charcoal and charred wood, on account of their durability, for landmarks in the ground between estates. The incorruptibility of charcoal is well known. Amongst other advantages, rats will not touch it; neither will the white ants nor cockroaches, so common in the Indies, commit their depredations where charring has been employed.

The 'Revue Horticole' states that it has been proved by recent experiments, that the best mode of prolonging the duration of wood is to char it, and then paint it over with three or four coats of pitch. Many of the sleepers now laid down on the Belgian railways are charred, the engineers preferring this process to any other.

The superficial carbonization, or charring of wood, as a preservative means, has long been practised. The Venetians have used charring for timber for a long period, particularly for piles. In France, M. de Lapparent recently proposed to apply it to the timber used in the French Navy. Some experiments, which were undertaken with a view to determine its practicability, terminated satisfactorily; and the Minister of Marine ordered the process to be introduced into the Imperial dockyards.

M. de Lapparent makes use of a gas blowpipe, the flame from which is allowed to play upon every part of the piece of timber in succession. By this means the degree of torrefaction may be regulated at will. The method is applicable to woodwork of all kinds; and the charring, it is said, does not destroy the sharpness of any mouldings with which the wood may be ornamented.

In the 'Journal des Savants,' Feb. 15, 1666, appears the following: "The Portugals scorch their ships, insomuch that in the quick works there is a coaly crust of about an inch thick; but this is dangerous, it happening, not seldom, that the whole vessel is burnt." It is no wonder that the Portuguese ships should frequently fire in the operation, as their plank was charred an inch deep. A mere charring, if done properly, after the timbers had been thoroughly seasoned by air, would have been sufficient.

Charring seasoned wood is known to be a most effectual mode of preservation against rot in timber: thus do piles, when charred, last for ages in water or moist soil. Charred wood has been dug up, which must have lain in the ground for 1500 years, and was then found perfectly sound. After the Temple of Diana, at Ephesus, was destroyed, it was found to have been built on charred piles; and at Herculaneum, after 2000 years, the charred wood was found to be whole and undiminished. But we find Sir Christopher Wren did not approve of charred piles, except in a soil where they would be constantly wet. So, in order to attain a firmer foundation for St. Paul's Cathedral, he had the ground excavated to an immense depth before a stone of the building was laid.

From time immemorial it has been the practice, particularly in France, to burn the ends of the poles driven into the ground to preserve them from decay. According to the remark of the celebrated Carlomb, we should always take into serious consideration old and well-known customs; but in this instance it is easy to admit the preserving effect of carbonization. Mr. James Randall,* Architect, states that he "oxidated several pieces of wood with nitric acid, and with fire," and these processes were attended with success. Nearly the last sentence in his work is, "*oxidation only* can be relied on, in all cases, as an effectual cure."

In charring, the surface of the timber is subjected to a considerable heat, the primary effect of which is to exhaust the sap of the epidermis, and to dry up the fermenting principles. Here this is done by long exposure to the air; and, in the second place, below the outside layer completely carbonized, a scorched surface is found, that is to say, partly distilled and impregnated with the products of that distillation, which is creosoted; the antiseptic properties of which are well known.

When Mr. Binmer was examined before the Commis-

* 'Directions to Cure the Dry Rot.' 1807.

sioners of Woods, Forests, &c., in 1792, he stated "that all steamed plank should be afterwards dried and *burnt* to extract the moisture."

To a spontaneous carbonization must be attributed also the unchangeableness of that timber entirely black, which is met with everywhere in digging up the ground, where it has laid buried for ages. In the neighbourhood of St. Malo, France, these specimens are very common, and there most of the espaliers and vine props are made of wood, black as ebony, and famous for its durability. They have been cut from the trees of an old forest, submerged in the eighth century by an inroad of the sea, which formerly crossed a Roman road, leading from Brittany to Cotentin.

Not long after the beginning of the eighteenth century, the method of heating or charring timber, before it was worked up, and also that of stoving—that is, of heating in kilns with sand—were practised in the Royal dockyards. The 'Royal William,' one of the most remarkable instances of durability that the British Navy has supplied, was built either wholly or in part of timber that had been charred. It was launched in 1719; never repaired until 1757; and then, when surveyed afloat, in 1785, it appeared that the thick stuff and plank had been *burnt* instead of being *kilned;* and that the ends of the beams, the faying parts of the breast-hooks, crutches, resters, knees, &c., had been gouged in a manner then practised, which was called *snail-creeping;* by means of which the air was conveyed to the different parts of the ship.*

* See Report of the Officers of Portsmouth Yard, 1792.

The reason this method has not been persevered in, but nearly abandoned, is owing to many causes: the difficulty and danger of the means adopted for charring, when either straw, fern, or shavings are made use of; the serious objection of burning the timber too deeply; or the encumbrance of the apparatus, and the length of time occupied, if sand-kilns sufficiently heated are used; and, finally, to indifference, or that system of routine, against which the wisest plans often contend in vain.

In house-building, the charring process should be applied to the beams and joists embedded in the walls, or surrounded with plaster; to the joists of stables, wash-houses, &c., which, although exposed to the free air, are constantly surrounded by a warm and moist atmosphere, an active cause of fermentation; to the wainscotting of ground floors; to the flooring beneath parquet work; to the joints of tongues and rabbets; for carbonization by means of gas still leaves to the wood, for working purposes, all the sharpness of its edges. Charring is particularly useful in the junction of all broad surfaces, and more essentially in those which are cut either transverse or oblique to the grain of the wood, as the sap vessels are then exposed to the absorption of moisture. The butts of timbers are peculiarly liable to rot, because of affording a lodgment for moisture without a free passage for air. No seasoned timber should have its tubular parts exposed, nor should any timber have the saw marks upon it, because the torn filaments absorb and retain moisture. Allusion has already been made to the process adopted, near Cherbourg, for preventing the decay of timber by means of gas.

By carbonization, a practical and economical means is afforded to railway companies of preserving, almost for ever, the sleepers, and particularly oak, which cannot be impregnated easily by the injection of mineral salts. Let us suppose, for instance, that after, say ten or fifteen years, the sleepers on a line are taken up for the length of a mile, and replaced by new ones; the old, when rasped and burnt again, will serve for the replacing the following mile, and so on, one mile after the other. It might be equally serviceable to apply the same process to injected beech, for the reason that it is almost impossible to make the preserving liquid penetrate thoroughly the mass of the timber.

SEASONING BY EXTRACTION OF SAP.

Mr. John Stephen Langton's method of seasoning by extraction of the sap was patented in 1825, but is now almost wholly discontinued. It consists in letting the timber into vertical iron cylinders, standing in a cistern of water, closing the cylinders at top; and the water being heated, and steam used to produce a partial vacuum, the sap relieved from the atmospheric pressure oozes from the wood, and being converted into vapour, passes off through a pipe provided for the purpose. The time required is about ten weeks, and the cost is about ten shillings per load; but the sap is wholly extracted, and the timber is said to be fit and ready for any purpose; the diminution of weight is, with a little more shrinkage, similar to that in seasoning by the common natural process.*

* See Tredgold's Report on this process, May 2, 1828.

Mr. Barlow's patent provided for exhausting the air from one end of the log while one or more atmospheres press upon the other end. This artificial aerial circulation through the wood is prolonged at pleasure. However excellent in theory, this process is not practicable.

In October, 1844, M. Tissier proposed to place wood in a close vessel, and subject it to a current of hot dry air; and in 1847, Mr. Miller proposed to inject hot air through beams of wood to drive out the sap.

In 1851, M. Meyer d'Uslaw proposed to first dilate the pores of the wood with steam, and then place it in a hermetically closed chamber, and make a vacuum there.

The following system of preparing timber for the Navy was, not many years since, adopted in South Russia. A full account of the practice will be found in Oliphant's 'Russian Shores of the Black Sea,' 1853. The only name we can give it is

"'SEASONING' BY BRIBES."

A certain quantity of well-seasoned oak being required, Government issues tenders for the supply of the requisite amount. A number of contractors submit their tenders to a board appointed for the purpose of receiving them, who are regulated in the choice of a contractor not by the amount of his tender, but of his bribe. The *fortunate* individual selected immediately sub-contracts upon a somewhat similar principle. Arranging to be supplied with the timber for half the amount of his tender, the sub-contractor carries on the game, and perhaps the eighth link in this contracting chain is the man who, for an absurdly low figure, undertakes to produce the *seasoned* wood.

His agents in the central provinces accordingly float a quantity of green pines and firs down the Dnieper and Bog to Nicholaeff, which are duly handed up to the head contractor, each man pocketing the difference between his contract and that of his neighbour. When the wood is produced before the board appointed to inspect it, another bribe *seasons* it; and the Government, after paying the price of well-seasoned oak, is surprised that the 120-gun ship, which it has been built of it, is unfit for service in five years.

> "Mark but my fall, and that that ruin'd me,
> Corruption."—SHAKSPEARE.

A few words can only be given to a most important matter, viz., the *second seasoning*, which many woods require. If floor-boards are only laid down at first on the joists of a building, and at the expiration of one year wedged tight and nailed down, those unsightly openings caused by shrinkage, which form a harbour for dirt and vermin, will be avoided, as the wood will have had an opportunity of shrinking. Doors, sashes, architraves in long lengths, will also be better if made up some time before they are required for use. Many Indian woods require a second seasoning—kara mardá, for instance, a favourite wood with Indian railway engineers. Even sál and teak are not exempt. Teak shrinks sideways least of all woods. In the 'Tortoise,' store ship, when fifty years old, no openings were found to exist between the boards; yet Colonel Lloyd says he found the teak timbers used by him in constructing a large room in the Mauritius to have shrunk ¾ of an inch in 38 feet. Thus a space of ⅜ of an inch must have been left at each end of the beam, where moisture could lodge and fungi exist, obtaining their nourishment from the wood. If unseasoned teak is used for ships, dry rot will in time find a place. It may be said that teak is a very hard wood, and very durable; yet "the mills of the gods," says an ancient philosopher, "grind slow, very slow, but they grind to powder;" *and so do the fungi mills.*

CHAPTER V.

ON SEASONING TIMBER BY PATENT PROCESSES, ETC.

Long years of practical experience has shown that timber, however prone to dry or wet rot, may be preserved from both by the use of certain metallic solutions, or other suitable protective matters.

All the various processes may be said somewhat to reduce the transverse strength of the timber when dry, and the metallic salts are affected at the iron bolts or fastenings. The natural juices of some woods do this; and bolts which have united beams of elm and pitch pine will often corrode entirely away at the junction.

The processes adopted for resisting the chemical changes in the tissues of the wood are all founded on the principle that it is essential to inject some material which shall at once precipitate the coaguable portion of the albumen retained in the tissues of the wood in a permanent insoluble form, so that it will not hereafter be susceptible of putrefactive decomposition. For this purpose, many substances, many solutions, have been employed with variable success, but materials have been sometimes introduced for this purpose which produced an effect just the opposite to what was anticipated.

Experience has shown that timber is permeable, at least by aqueous solutions, only so long as the sap channels are free from incrustation.

Such in general is the case with beech, elm, poplar, and hornbeam, the capillary tubes of which are always open, or, at least, close very slowly. At the same time it may be said that there must remain ever in these species some parts impervious to injection, whilst it is almost impossible but that a certain portion of the fibres will be more or less incrusted. The sap woods, on the other hand, of every species appear quite pervious.

Very little is known of any preservative process adopted in ancient times. Pliny observes that the ancients used garlic boiled in vinegar with considerable success, especially with reference to preserving timber from worms: he also states that the oil of cedar will protect any timber anointed with it from worm and rottenness. Oil of cedar was used by the ancient Egyptians for preserving their mummies. Tar and linseed oil were also recommended by him. The image of the goddess Diana, at Ephesus, was saturated with olive and cedar oils; also the image of Jupiter, at Rome; and the statues of Minerva and Bacchus were impregnated with oil of spikenard.

The idea of preserving wood by the action of oil is therefore by no means new; but it is somewhat curious that the earliest modern processes should also be by means of oil. The *oils most proper* to be used are *linseed*, *rapeseed*, or almost any of the vegetable fixed oils. Oak wood, rendered entirely free from moisture, and then immersed in linseed oil, is said to be thus prevented from splitting: the time of immersion depending on the size, &c. Palm oil is preferable to whale oil, because impregnation with the latter, although in many instances eligible, causes wood to become brittle. It is, however, probable that whale oil, when combined with other substances, such as litharge, coal pitch, or charcoal, may lose much of that effect. As cocoa-nut oil, which is, under low temperature, like the oil expressed from the nuts of the palm tree, is known to be highly preservative of timber and metallic fastenings, we may expect the same result from the latter, and thereby avoid that extreme dryness and brittleness of the timber which Mr. Strange complained of in the Venetian ships that had been seasoned for many years in frame under cover. Cocoa-nut oil beat up with shell lime or *chunam*, so as to become putty, and afterwards diluted with more oil, is used at Bombay and elsewhere as a preservative coat or varnish to plank. It cannot become a varnish without the addition of some essential oil; and the oil of mustard is used; which, of course, will produce the desired effect. In the first volume of the Abbé Raynal, on the European settlements in the East and West Indies, he mentions that an oil was exported from Pegu for the preservation of ships; but as he does not say what oil, no conclusion can be drawn further than as to the probability of its being one of those already noticed.

Experience has proved that even *animal oils* are so far injurious to timber as to *render it brittle*, whilst they preserve it from rottenness; and that, on the other hand, a mineral salt more or less combined with fatty substances does not produce that effect. The staves of whale-oil casks become quite brittle, whilst those of beef, pork, and tallow barrels remain tough and sound. Ships *constantly* in the Greenland trade have their timbers and planks preserved so far as they have become impregnated with whale oil.

Experiments with fish oil prove that of itself, unless exposed to sun and air, it may be injurious; that it loosens the cohesion of timber; but that *animal fat, combined with saline matter, is preservative.*

Fish oil used alone is ineligible, because capable of running into the putrefactive process, unless as a thin outside varnish. In hard, sound timber, it will hardly enter at all; and if poured into bore-holes in the heads of timbers, it will insinuate itself into the smallest rents or cracks, and waste through them. Used alone, or with any admixture, it is absorbed and dried quickly on wood in a decomposing state or commencing to be dry rotten. Used with litharge, it dries after some days; but with lamp-black it has scarcely so much tendency to dry as when used alone. Paint of fish oil and charcoal dries very quickly where there is absorption, and the charcoal extends its oxidating or drying effect to the fish oil in its vicinity.

We give the following to prove what we have written, and also to serve as an example for those who wish to try experiments:—

EXPERIMENTS ON FISH OIL.

June 9.—Upon a piece of old oak scantling, with its alburnam on one side in a state of decay, fish oil was poured several times, viz. on this day, on June 25, and July 3, which it rapidly absorbed in the decayed part.

July 26.—It was payed (or mopped) with fish oil and charcoal powder, and the following day it was put under an inverted cask.

October 1.—The end of this piece was covered with a greenish mould. *This proves that fish oil must be injurious, except where exposed to sun and air to dry it.*

A compound of fixed oils and charcoal is liable to inflame, but as a thin covering or pigment it may not be so.

The *petroleum* oil-wells, near Prome, in Burmah, have been in use from time immemorial. Wood, both for ship-building and house-building, is invariably saturated or coated with the product of those wells; and it is stated that the result is *entire immunity from decay* and the ravages of the white ant. At Marseilles, and some other ports in the Mediterranean, it used to be the practice to run the petroleum, which is obtained near the banks of the Rhone, into the vacancies between the timbers of the vessels, to give them durability. It was sometimes, for the conjoint purpose of giving stability and duration to vessels, mixed with coarse sand or other extraneous matters, and run in whilst hot between the ceiling and bottom plank, where it filled up the vacancies between the timbers in the round of their bottoms, excepting where necessary to be prevented. The great objection to the use of petroleum is its inflammability. *Creosote*, its great rival for wood preserving, is also inflammable, and not so agreeable in colour; but it is *considerably cheaper*, which is an important matter.

As we are now about to enter upon the subject of patent processes, &c., it appears desirable to lay down certain principles at the commencement, in order to assist the reader as much as possible.

Almost every chemical principle or compound of any plausibility has been suggested in the course of the last hundred and fifty years; but the multiplicity and contradiction of opinions form nearly an inextricable labyrinth. To commence.

1st. It seems obvious that *the sooner the sap is wholly removed from the wood the better, provided the woody fibre solidifies without injury.*

2nd. That *the wood should be impregnated with any strongly antiseptic and non-deliquescent matter, which must necessarily be in solution when it enters the wood.* No deliquescent remedy is eligible, because moisture is injurious to metallic fastenings.

3rd. *The wood should be first dried, and its pores then closed with any substance impervious to air and moisture, and at the same time highly repellant to putrescency.* The most essential requisites in a preservative of timber being a disposition to *dryness*, and a tendency to resist *combustion* as far as consistently obtainable.

4th. *Any process to be successful ought not to be tedious, very difficult, or too expensive.* These are important elements in the success of any patent.

Very little is known of any preservative process previous to the year 1717, when directions were given by the Navy authorities to *boil* treenails, and dry them before they were used. But whether the custom had prevailed before this time, or whether their strength and durability were increased by it, there are no means of ascertaining. It does not appear that any substance was put into the water to decompose the juices; but as they are soluble in warm water, perhaps the power of vegetation might have been destroyed without it.

In 1737 Mr. Emerson patented a process of saturating

timber with *boiled oil*, mixed with poisonous substances; but his process was very little used. This, we believe, was the *first* patent on wood preserving.

About 1740, Mr. Reid proposed to arrest decay by means of a certain *vegetable acid* (probably pyroligneous acid). The method of using it was by simple immersion.

In 1756, Dr. Hales recommended that the planks at the water-line of ships should be soaked in *linseed oil*, to prevent the injury to which wood is subject when alternately exposed to wet and dry; and indeed, many ships were built in which a hollow place was cut in one end of each beam or sternpost, which might constantly be kept filled with *train oil*. Amongst other ships so constructed, the 'Fame,' 74, may be mentioned. When, after some years, this ship was repaired, it was found that as far as the oil had penetrated, namely, from 12 to 18 inches from the end, the wood was quite sound, whilst the other parts were more or less decayed. The Americans used to hollow out the tops of their masts in the form of cups or basins; bore holes from the end a considerable way down the masts; pour oil into these; cover them over with lead; and leave the oil to find its way down the capillary vessels to the interior of the timber.*

In 1769, Mr. Jackson, a London chemist, with a view to the prevention of decay, obtained permission to prepare some timber to be used in the national yards, by immersing it in a solution of *salt water, lime, muriate of soda, potash, salts*, &c., the result of which dose was, that several frigates in the Navy subjected to the process were rendered

* See No. 1, p. 3, Appendix to first volume of 'Naval Architecture.'

more perishable than if they had been constructed of unprepared timber. The solution was filtered into the wood partly by means of holes made in it. Chapman proposed a similar method of preserving the frames of ships, viz. by boring holes in the timbers, and pumping a *solution of copperas* in water into them. He believed every part of the vessel would thus be impregnated.

Mr. Jackson also prepared the frame of the ship 'Intrepid' with another solution. The ship lasted many years. Bowden thought it was a *solution of glue*. Chapman suggested *slaked lime*, thinned with a weak *solution of glue* for mopping the timbers of a ship.

Shortly after Mr. Jackson's process was started, Mr. Lewis attempted to accomplish the preservation of timber by placing it surrounded by pounded lime, in spaces below the " surface of the earth." The use of lime has also been advocated by Mr. Knowles, Secretary of the Committee of Surveyors of the Navy, who has written an able work on the 'Means to be taken to Preserve the British Navy from Dry Rot' (1821).

Between 1768 and 1773 a practice prevailed of saturating ships with common salt; but this was found to cause a rapid corrosion of the iron fastenings, and to fill the vessels between decks with a constant damp vapour. In 'Nicholson's Journal,' No. 30, there is an article signed *Nauticus* on this subject. Vessel owners had long ago observed that those ships which have early sailed with cargoes of salt are not attacked by dry rot. Indeed, several instances are attested of vessels whose interiors were lined with fungi having all traces of the plant destroyed by accidental or intentional sinking in the sea. Acting on such hints, a trader of Boston, U. S., salted his ships with 500 bushels of the chloride, disposed as an interior lining, adding 100 bushels at the end of two years. Such an addition of dead weight is sufficient objection to a procedure which has other great disadvantages. Salt should never be applied as an antidote against the dry rot, on account of its natural powers of attracting moisture from the atmosphere, which would render apartments almost uninhabitable, from their continual dampness. Those who have lived for any length of time in a house at the sea-side, the mortar of which has been partly composed of sea sand, will have observed the moist state of the paper, plastering, &c., in wet weather. Bricks made with sea sand are objectionable.

Salt water seasoning has already been referred to in the last chapter, but as it is so closely connected with salt seasoning, the further and final consideration of salt water seasoning may be fitly dealt with here. Salt water will not extract the juices from the timber like fresh water. It is only by destroying the vegetation that salt water can be advantageous, but it would require a very long time to impregnate large timber to the heart so as to destroy vegetation. It is well known that wood is softened, and in time decomposed, by extreme moisture. Fifty years since, the master builder at Cronstadt complained that the oak from Casan, which was frequently wet from different causes in its passage of three years to Cronstadt, was so water-soaked as never to dry; and also from the information of Mr. Strange, it appears "that the practice at Venice of the fresh-cut timber being thrown into salt water, prevents its ever becoming dry in the ships, and that the salt water rusted and corroded the iron bolts." In fine, vessels built with salt water seasoned wood are perfect hygrometers, being as sensible to the changes of the moisture of the atmosphere as lumps of rock salt, or the plaster of inside walls where sea sand has been used.

In Ceylon, the timber of the female palm tree is much harder and blacker than that of the male, inasmuch as it brings nearly triple its price. The natives are so well aware of the difference that they resort to the devise of immersing the male tree in *salt water* to deepen its colour, as well as add to its weight.

Vessels impregnated with *bay salt*, or the large grained salt of Leamington or of Liverpool (pure muriate of soda), will possess decided advantages; as also will vessels that have been laden with *saltpetre*, if it has been dispersed amongst their timbers.

Ships (the timbers of which had been previously immersed in salt water) have been broken up after a few years' service, and the floor timbers taken out quite sound: but when exposed to the sun and rain in the summer months, their albumen has been in a decomposed or friable state.

By the answers to queries given to Mr. Strange, the British Minister at Venice, in or about 1792, it appears that several of the Venetian ships of war had then lain under sheds for fifty-nine years; some in bare frames, and others planked and caulked: that these ships show no outward marks of decay; but their timbers have shrunk much, and *become brittle;* that some of the most intelligent ship builders were of opinion that great prejudice had arisen from the prevalent custom of throwing the timber fresh cut into salt water, and letting it lie there until wanted; that afterwards it dried, and withered on the outside, under the sheds, while the inside, being soaked with salt water, rotted before it became dry; and this was one reason, amongst others, why Venetian ships, though built of good timber, lasted so short a time; for the salt moisture not only rots the inside of the beams and timbers, but of course rusts and corrodes the iron bolts.

Salt water, sea-sand, and *sea-weed* are now used for seasoning "jarrah" wood in Western Australia. This wood is considered a first-class wood for shipbuilding, but it is somewhat slow to season, and if exposed before being seasoned it is apt to "fly" and cast. The method adopted is as follows: The logs are thrown into the sea, and left there for a few weeks; they are then drawn up through the sand, and after being covered with sea-weed a few inches deep, are left to lie on the beach, care being taken to prevent the sun getting at their ends. The logs are then left for many months to season. When taken up they are cut into boards 7 inches wide, and stacked, so as to admit of a free circulation of air round them, for five or six months before using them. Sea-weed or sea-ware, cast upon the shores, contains a small quantity of carbonate of soda, and a large proportion of nitrogenous and saline matters, with earthy salts, in a readily decomposable state. They also contain much soluble mucilage. The practice of seasoning timber by heating it in a *sand bath* was formerly adopted by the Dutch, and by the Russians in building boats.

Mr. Thomas **Nichols** (in a letter to Lord Chatham, when First Lord of the Admiralty) states "that the same end, viz. preservation of timber from decay, might probably be acquired by burying the timber in sand, which acts as an artificial sap," in the same manner as mentioned in Townsend's 'Travels through Spain,' to be used with the masts of ships of war at Cadiz.

Peat moss has been recommended (because the sulphates of iron, soda, and magnesia are found in it), but it failed when tried.

With reference to Mr. Lewis's proposal to preserve wood by means of lime, it must be remembered that *quicklime*, with damp, has been found to accelerate putrefaction, in consequence of its extracting carbon; but when dry, and in such large quantities as to absorb all moisture from the wood, *the wood is preserved, and the sap hardened*. Vessels *long* in the lime trade have afforded proof of this fact; and we have also examples in plastering-laths, which are generally found sound and good in places where they have been dry. Whitewash or *limewater* has been strongly recommended for use between the decks of ships, as being unfavourable to vegetation: it should be renewed at intervals of time, according to circumstances. It has been applied with good effect to the joists and sleepers of kitchen floors; but to be effectual it should be occasionally renewed. Effete, or re-carbonated lime, is injurious to timber, like other absorbent earths; so also are calcareous incrustations formed by the solution of lime in water, as appears from Von Buch's 'Travels in Norway,' in which he says, "that in the fishing country (near Lofodden, beyond the Arctic circle) the calcareous incrustations brought by water, filtering through a bed of shells, soon cause the vessels and wood to be covered with and destroyed by green fungi." The ends of joists of timber inserted in walls are frequently found rotten; and where not so, it may probably be owing to the mortar having been made with hot lime, and used immediately, or to the absence of moisture. It does not appear practicable to use limewater to any extent for preserving timber, because water holds in solution only about $\frac{1}{500}$th part of lime, which quantity would be too inconsiderable; it, however, renders timber more durable, but at the same time very hard and difficult to be worked (p. 73).

Vessels constantly in the *coal* trade have generally required little repair, and have lasted until in the common course of things they were lost by shipwreck. This must be owing to the martial pyrites which abound in all coals; and also from the sulphuric acid arising from the quantity of coal dust which finds its way through the seams of the ceiling, and adheres to the timber and planks.

In 1779, M. Pallas, in Russia, proposed to steep wood in *sulphate of iron* (green vitriol) until it had penetrated deeply, *and* then in *lime* to precipitate the vitriol. Neumann, in his first volume of 'Chemistry,' on the article green vitriol, says, "That in the Swedish transactions this salt is recommended for preserving wood, particularly the wheels of carriages, from decay.

"When all the pieces are fit for being joined together, they are directed to be boiled in a solution of vitriol for three or four hours, and then kept for some days in a warm place to dry. It is said that the wood by this preparation becomes so hard and compact that moisture cannot penetrate it, and that iron nails are not so apt to be destroyed in this vitriolated wood as might be expected, *but last as long as the wood itself.*"

In 1780 the marcasite termed by the miners *mundic*, found in great abundance in the tin mines in Devonshire and Cornwall, was employed, in a state of fusion, to eradicate present and to prevent the future growth of dry rot; but whether its efficacy was proved by time is not known. A garden walk where there are some pieces of mundic never has any weeds growing; the rain that falls becomes impregnated with its qualities, and in flowing through the walk prevents vegetation.

In 1796 Hales proposed to *creosote* the treenails of ships: this was forty-two years previous to Bethell's patent for creosoting wood.

About the year 1800, the Society of Arts' building in the Adelphi, London, being attacked by dry rot, Dr. Higgins examined the timbers, caused some to be removed and replaced by new, and the remainder to be scraped and washed with a solution of *caustic ammonia,* so as by burning the surface of the wood to prevent the growth of fungi.

At the commencement of the present century, a member of the Royal Academy of Stockholm called attention to the use of *alum* for preserving wood from fire. He says, in the Memoirs of that Academy, "Having been within these few years to visit the alum mines of Loswers, in the province of Calmar, I took notice of some attempts made to burn the old staves of tubs and pails that had been used for the alum works. For this purpose they were thrown into the furnace, but those pieces of wood which had been *penetrated* by the alum did not burn, though they remained for a long time in the fire, where they only became red; however, at last they were consumed by the intenseness of the heat, but they yielded no flame." He concludes, from this experiment, that wood or timber for the purpose of building may be secured against the action of fire by letting it remain for some time in water wherein vitriol, alum, or any other salt has been dissolved which contains no inflammable parts.

In Sir John Pringle's Tables of the antiseptic powers of different substances, he states *alum* to be thirty times stronger than sea-salt; and by the experiments of the author of the 'Essai pour servir à l'Histoire de la Putréfaction,' metallic salts are much more antiseptic than those with earthy bases.

In 1815 it occurred to Mr. Wade that it would be a good practice to fill the pores of timber with *alumine*, or selenite; but two years after, Chapman observed, " Impregnation of ships' timbers with a *solution of alum* occurred to me about twenty years since, because on immersion in sea-water the alumine would be deposited in the pores of the timber; but I was soon informed of its worse than inutility, by learning that the *experiment had been tried*, and, in place of preserving, had *caused the wood to rot speedily*. Impregnation with *selenite* has been tried in elm water-pipes. On precipitation from its solvent it partially filled the pores, and hardened the wood, but *occasioned speedy rottenness.*" If, by using a *solution of alum* to render wood uninflammable, we at the same time cause it to *rot speedily*, it becomes a question *whether the remedy is not worse than the disease.* Captain E. M. Shaw, of the London Fire Brigade, in his work, 'Fire Surveys' (1872), recommends *alum* **and water.** Probably he only thought of fire, and not of rotting the wood. The alum question does not appear to be yet satisfactorily settled.

While upon the subject of uninflammable wood, we may state that in 1848, upon Putney Heath (near London), by the roadside, stood an obelisk, to record the success of a discovery made in the last century of the means of building a house which no ordinary application of ignited combustibles could be made to consume: the obelisk was erected in 1786. The inventor was Mr. David Hartley, to whom the House of Commons voted 2500*l.*, to defray the expenses of the experimental building, which stood about one hundred yards from the obelisk. The building was three stories high, and two rooms on a floor. In 1774, King George the Third and Queen Charlotte took their breakfast in one of the rooms, while in the apartment beneath fires were lighted on the floor, and various inflammable materials were ignited to attest that the rooms above were fire-proof. Hartley's secret lay in the floors being double, and there being interposed between the two boards sheets of laminated iron and copper, not thicker than stout paper, which rendered the floor air-tight and thereby intercepted the ascent of the heated air; so that, although the inferior boards were actually charred, the metal prevented the combustion taking place in the upper flooring. Six experiments were made by Mr. Hartley in this house in 1776, but we cannot ascertain any particulars about them, or any advantages which accrued to the public from the invention, although the Court of Common Council awarded him the freedom of the City of London for his successful experiments.

In 1805 Mr. Maconochie proposed to saturate with resinous and oily matters inferior woods, and thus render them more lasting. This proposal was practically carried out in 1811 by Mr. Lukin, who constructed a peculiar stove for the purpose of thus impregnating wood under the influence of an increased temperature. The scheme, however, had but very partial success, for either the heat was too low and the wood was not thoroughly aired and seasoned, or it was too high and the wood was more or less scorched and burnt. Mr. Lukin buried wood in *pulverized charcoal* in a heated oven, but the fibres were afterwards discovered to have started from each other. He next erected a large kiln in Woolwich dockyard, capable of containing 250 loads of timber, but an explosion took place on the first trial, before the process was complete, which proved fatal to six of the workmen, and wounded fourteen, two of whom shortly afterwards died. The explosion was like the shock of an earthquake. It demolished the wall of the dockyard, part of which was thrown to the distance of 250 feet; an iron door weighing 280 lb. was driven to the distance of 230 feet; and other parts of the building were borne in the air upwards of 300 feet. The experiment was not repeated.

Mr. Lukin was not so fortunate in 1811 as in 1808, for in the latter year he received a considerable reward from the Government for what was considered a successful principle of ventilating hospital ships.

In 1815 Mr. Wade recommended the impregnation of timber with resinous or oleaginous matter (preferring linseed oil to whale oil) or with *common resin* dissolved in a lixivium of *caustic alkali*, and that the timber should afterwards be plunged into water acidulated with any cheap acid, or with alum in solution. He considered that timber impregnated with oil would not be disagreeable to rats, worms, cockroaches, &c., and that the contrary was the case with resin. He also recommended the impregnation of timber with *sulphate of copper, zinc, or iron*, rejecting deliquescent salts, as they corrode metals.

In 1815 Mr. Ambrose Boydon, of the Navy Office, strongly recommended that the timber, planks, and treenails of ships should be first boiled in *limewater* to correct the acid, and that they afterwards should be boiled in a *thin solution of glue*, by which means the pores of the wood would be filled with a hard substance insoluble by water, which would not only give the timbers durability, by preventing vegetation, but increase their strength. Glue, he thought, might be used without limewater, or glue and limewater mixed together.

In 1817 Mr. William Chapman published the result of various experiments he had made on wood with *lime, soap,* and *alkaline* and *mineral salts*. He recommended a solution of a pound of *sulphate of copper* or blue vitriol (at that time 7*d*. per pound) dissolved in four ale gallons of rain water, and mopped on hot over all the infected parts, or thrown over them in a plentiful libation. He also recommended one ounce of *corrosive sublimate* (then 6s. per pound) to a gallon of rainwater applied in the same manner to the infected parts. For weather-boarded buildings he considered one or more coats of thin *coal tar*, combined with a small portion of *palm oil*, for the purpose of preventing their tendency to rend, to be a good preservative.

Messrs. Wade, Boydon, and Chapman published works on dry rot about this time.

In 1822 Mr. Oxford took out a patent for an improved method of preventing "decay of timber," &c. The process proposed was as follows: "The essential *oil of tar* was first extracted by distillation, and at the same time saturated with *chlorine gas*. Proportions of *oxide of lead*, *carbonate of lime*, and *carbon of purified coal tar* well ground, were mixed with the oil, and the composition was then applied in thick coatings to the substances intended to be preserved.

On 31st March, 1832, Mr. Kyan patented his process of *corrosive sublimate* (solution of the *bi-chloride of mercury*) for preventing dry rot; which process consisted as follows: A solution of the corrosive sublimate is first made, and the timber is placed in the tank. The wood is held down in such a way, that when immersed on the fluid being pumped in, it cannot rise, but is kept under the surface, there being beams to retain it in its place. There it is left for a week, after which the liquor is pumped off, and the wood is removed. This being done, the timber is dried, and said to be prepared. Sir Robert Smirke was one of the first to use timber prepared by Kyan, in some buildings in the Temple,

London; and he made some experiments on timber which had undergone Kyan's process. He says, "I took a certain number of pieces of wood cut from the same log of yellow pine, from poplar, and from the common Scotch fir; these pieces I placed first in a cesspool, into which the waters of the common sewers discharged themselves; they remained there six months; they were then removed from thence, and placed in a hotbed of compost, under a garden-frame; they remained there a second six months; they were afterwards put into a flower border, placed half out of the ground, and I gave my gardener directions to water them whenever he watered the flowers; they remained there a similar period of six months. I put them afterwards into a cellar where there was some dampness, and the air completely excluded; they remained there a fourth period of six months, and were afterwards put into a very wet cellar. Those pieces of wood which underwent Kyan's process are in the same state as when I first had them, and all the others to which the process had not been applied are more or less rotten, and the poplar is wholly destroyed.

"I applied Kyan's process to yellow Canadian pine about three years ago, and exposed that wood to the severest tests I could apply, and it remains uninjured, when any other timber (oak or Baltic wood) would certainly have decayed if exposed to the same trial, and not prepared in that manner.

"As another example of the effect of the process, I may mention that about two years ago, in a basement story of some chambers in the Temple, London, the wood flooring and the wood lining of the walls were entirely decayed from the dampness of the ground and walls, and to repair it under such circumstances was useless. As I found it extremely difficult to prevent the dampness, I recommended lining the walls and the floor with this prepared wood, which was done; and about six weeks ago I took down part of it to examine whether any of the wood was injured, but it was found in as good a state as when first put up. I did not find the nails more liable to rust.'

"I have used Kyan's process in a very considerable quantity of paling nearly three years ago; that paling is now in quite as good a state as it was, though it is partly in the ground. It is yellow pine. Some that I put up the year before, without using Kyan's process (yellow pine), not fixed in the ground, but close upon it, is decayed."

This evidence, by such an experienced architect as the late Sir Robert Smirke was, is certainly of great value in favour of Kyan's process.

The recorded evidence upon the efficiency of this mode of treating timber for its preservation is somewhat contradictory. On the Great Western Railway 40,000 loads were prepared, at an expenditure of $1\frac{3}{4}$ lb. of sublimate to each load, the timber, 7 inch, being immersed for a period of eight days, and the uniformity of the strength of the solution being constantly maintained by pumping. Some samples of this timber, after six years' use as sleepers on the railway, were found "as sound as on the day on which they were first put down." This timber was prepared by simple immersion only, without exhaustion or pressure. Some of the sleepers on the London and Birmingham Railway, on the other hand, which had been Kyanized three years only, were found absolutely rotten, and Kyan's process was there consequently abandoned.

This process is said to cost an additional expense to the owner of from fifteen to twenty shillings per load of timber. Mr. Kyan at first used 1 lb. of the salt in 4 gallons of water, but it was found that the wood absorbed 4 or 5 lb. of this salt per load; more water was added to lessen the expense, until the solution became so weak as in a great measure to lose its effect.

Simple immersion being found imperfect as a means of injecting the sublimate, attempts were afterwards made to improve the efficiency of the solution by *forcing* it into the wood. Closed tanks were substituted for the open ones, and forcing pumps, &c., were added to the apparatus. The pressure applied equalled 100 lb. on the square inch. With this arrangement a solution was made use of having 1 lb. of the sublimate to 2 gallons of water; and it was found that three-fourths of this quantity sufficed for preparing one load of timber. The timber was afterwards tested, and it was ascertained that the solution had penetrated to the heart of the logs. Mr. Thompson, the Secretary to Kyan's Company, stated, in March, 1842, that experience had proved "that the strength of the mixture should not be less than 1 lb. of sublimate to 15 gallons of water; and he had never found any well-authenticated instance of timber decaying when it had been properly prepared at that strength." As much

PATENT PRESERVATIVE SYSTEM.

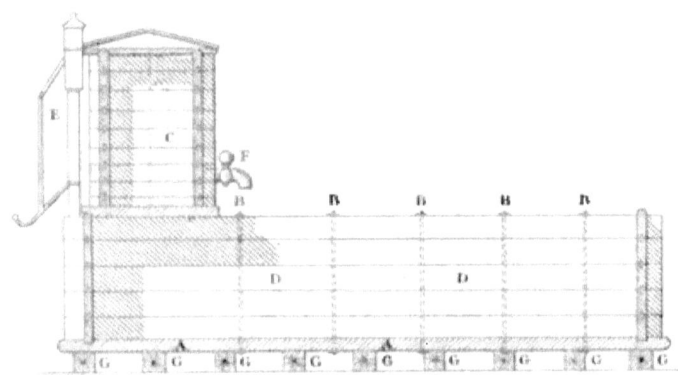

Horizontal Section of Mr. Kyan's Original Tank and Cistern.

A. Bottom of Tank
B. ¾ Iron bolts to connect the planks which form the sides and end of the Tank and Cistern.
C. The Cistern which contains the solution.
D. The Tank.
E. Pump for raising the solution from Tank into Cistern
F. Tap for conveying the solution from Cistern to Tank.
G. Wood sleepers to carry Tank and Cistern.

as 1 in 9 was not unfrequently used. Kyan's process is now but very rarely used; Messrs. Bethell, of King William Street, London, adopt it when requested by their customers. We have given the statements which have been made for and against this patent, but after a lapse of forty years it is difficult to reconcile conflicting statements.

Although Mr. Kyan invented his process in 1832, Sir Humphrey Davy had previously used and recommended to the Admiralty, and Navy Board, a weak solution of the same thing to be used as a wash where rot made its appearance: on giving his opinion upon Mr. Lukin's process, that eminent chemist observed, "that he had found corrosive sublimate highly antiseptic, and preservative of animal and vegetable substances, and therefore recommended rubbing the surface of timber with a solution of it." In 1821 Mr. Knowles, of the Navy Office, referred to the use of corrosive sublimate for timber. In fact, it was used in 1705, in Provence (France), for preserving wood from beetles. Kyan, however, was the first to apply it to any extent. In the years 1833 to 1836, at the Arsenal, Woolwich, experiments were instituted, having for their object the establishing, or otherwise, the claims of Kyan's system; the results of which were of a satisfactory nature. Dr. Faraday has stated that the combination of the materials used was not simply mechanical but chemical; and Captain Alderson, C.E., having experimented upon some specimens of ash and Christiana deal, found that the rigidity of the timber was enhanced, but its strength was in some measure impaired; its specific gravity being also in some degree diminished.* Kyan's process is said by some to render the wood brittle.

Mr. Kyan considered that the commencement of rot might be stopped or prevented by the application of corrosive sublimate, in consequence of the chemical combination which takes place between the corrosive sublimate and those albuminous particles which Berzelius and others of the highest authority consider to exist in and form the essence of wood; which, being the first parts to run to decay, cause others to decay with them. By seasoning timber in the ordinary way, the destructive principle is dried, and under common circumstances rendered inert. But when the timber is afterwards exposed to great moisture, &c. (the fermentative principle being soluble when merely dried), it will sometimes be again called into action. Kyan's process is said not only altogether to destroy this principle and render it inert, but, by making it solid and perfectly insoluble, to remove it from the action of moisture altogether. It thus loses its hygrometric properties, and, therefore, prepared or patent seasoned timber is not liable to those changes of atmosphere which affect that which is seasoned in the common way. All woods, including mahogany and the finest and most expensive wood, may be seasoned by Kyan's process in a very short space of time, instead of the months required by the ordinary methods.

The reader will find a great deal about Kyan's system in the 'Quarterly Review,' April, 1833; and about pro-

---

* See paper on "Kyan's Process," by Captain R. C. Alderson, C.E., in vol. i. 'Papers of Royal Engineers.'

posals for using chloride of mercury for wood, 'Memoirs of the Academy of Dijon,' 1767; 'Bull. des Sciences tecn.,' v. ii., 1824, Paris; and 'Bull. de Pharm.,' v. 6, 1814, Paris.

It is well known that Canadian timber is much more liable to decay than that grown in the northern parts of Europe, and for this reason is never extensively used in buildings of a superior description. The principle of decay being destroyed by Kyan's process as above described, this objection no longer exists, and this kind of timber may therefore now be employed with as great security as that of a superior quality and higher price. The same observation applies with great force to timber of British growth, particularly to that of Scotland, much of which is considered as of little or no value for durable purposes, on account of its extreme liability to decay, whether in exposed situations or otherwise. The process invented by Kyan might therefore render of considerable value plantations of larch, firs of all kinds, birch, elm, beech, ash, poplar, &c.

Cost of process in 1832, 1*l.* per load of 50 cubic feet of timber.

Mr. W. Inwood, the architect of St. Pancras Church, London, reported favourably of Kyan's process. On 22nd February, 1833, Professor Faraday delivered a lecture at the Royal Institution, London, on Kyanizing timber; and on 17th April, 1837, he reported that Kyan's process had not caused any rusting or oxidation of the iron in the ship 'Samuel Enderby,' after the ship had been subjected to this process, and had been on a three years' voyage to the South Sea fisheries; and in the same year, viz. 1837, Dr. Dickson delivered a lecture at the Royal Institute of British Architects on dry rot, recommending Kyan's process.

Five years after Mr. Kyan's invention, viz. in 1837, a Mr. Flocton invented a process for preventing decay, by saturating timber with **wood-tar** and *acetate of iron*, but little is known of this invention: we believe it was a failure.

During the same year Mr. Flocton's process was made known, a Frenchman named Letellier recommended saturating timber in a solution of *corrosive sublimate*, and when dry, into one of *glue, size*, &c.*

During this year Mr. Margary took out his patent for applying *sulphate of copper* to wood. We propose to describe Margary's process further on: we do not think he received any medals for it.

We now arrive at the modern *creosoting* process, which was brought to perfection by the late Mr. John Bethell. Mr. Bethell's process of creosoting, or the injection of the heavy oil of tar, was first patented by him on July 11th, 1838.† It consists in impregnating the wood throughout with oil of tar, and other bituminous matters containing creosote, and also with pyrolignite of iron, which holds more creosote in solution than any other watery menstruum. Creosote, now so extensively used in preserving wood, is obtained from coal tar, which, when submitted to distillation, is found to consist of pitch, essential oil

---

* See Chapman, Boydon, Jackson, and Kyan's methods.

† See 'London Journal of Arts,' March, 1842; 'Bull. de l'Encouragement,' June, 1842.

(creosote), naphtha, ammonia, &c. In the application of the oil of tar for this purpose, it is now considered to be indispensable that the ammonia be got rid of; otherwise the wood sometimes becomes brown and decays, as may be constantly seen in wood coated with the common oil tar. The kind of creosote preferred by continental engineers and chemists, and also by the late Mr. John Bethell himself, is *thick,* and rich in *naphthaline.* Some English chemists now seem to prefer the thinnest oil, which contains no naphthaline, but a little more carbolic acid; the crude carbolic acid would vary from 5 to 15 per cent.: no engineer has ever required more than 5 per cent. of crude carbolic acid in creosote. The thinner oil appears to be more likely to be drawn out of the wood by the heat of the sun or absorption in powdery soil, and is more readily dissolved out by moisture.

Mummies many thousands of years old have evidently been preserved on the creosoting principle, and from observing the mummies the process of creosoting suggested itself to Mr. Bethell. The ancient Egyptians, whether from the peculiarity of their religious opinions, or from the desire to shun destruction and gain perpetuity even for their dead bodies, prepared the corpses of their deceased friends in a particular way, viz. by coagulating the albumen of the various fluids of the body by means of creosote, cedar oil, salt, and other substances, and also by excluding the air. How perfectly this method has preserved them the occasional opening of a mummy permits us to see. A good account of the operation is given in the chapter on mummies, in the second volume of Egyptian Antiquities in the 'Library of Entertaining Knowledge.'

By the process of creosoting the timber is rendered more durable, and less liable to the attack of worms; but it becomes very inflammable; that is, *when* once alight burns quickly; in addition to which, the disagreeable odour from timber so treated renders it objectionable for being used in the building of dwelling-houses.

The action of the solutions in water of metallic salts is, if the mixture is sufficiently strong, to coagulate the albumen in the sap; but the fibre is left unprotected.

Creosote has the same effect of coagulating the albumen, whilst it fills the pores of the wood with a bituminous asphaltic substance, which gives a waterproof covering to the fibre, prevents the absorption of water, and is obnoxious to animal life.

In cases where the complete preservation of timber is of vital importance, and expense not a consideration, the wood should be first subjected to Burnett's process, and then creosoted; by which means it would be nearly indestructible; the reason for this combined process being, that the albumen or sap absorbs the creosote more readily than the heart of the timber, which can, however, be penetrated by the solution of chloride of zinc. Mr. John Bethell's patent of 1853 recommends this in a rather improved form. He says the timber should *first* be injected with metallic salts, then dried in a drying-house, then creosoted. By this method, very considerable quantities both of metallic salt and creosote can be injected into timber.

It has been stated that the elasticity of wood is increased by creosoting; the heartwood only decays by oxidation.

The wood should be dried previous to undergoing the process, as the sapwood, otherwise almost useless, can be rendered serviceable, and for piles for marine work whole round timber should be used, because the sapwood is so much more readily saturated with the oil, and this prevents the worms from making an inroad into the heart.

Mr. Bethell uses about 10 lb. of creosote per cubic foot of wood, and he does not allow a piece of timber to be sent from his works without being tested to ascertain if it has absorbed that amount, or an amount previously agreed upon. We mention the latter statement, because it is evident that all descriptions of wood cannot be made to imbibe the same amount. This process is chiefly used for pine timber: yellow pine should absorb about 11 lb. to the cubic foot, and Riga pine about 9 lb. The quantity of oil recommended by the patentee, engineers, and others, is from 8 to 10 lb. for land purposes, and about 12 lb. to the cubic foot for marine. In this country, for marine the quantity does not exceed 12 lb.; but on the Continent, in France, Belgium, and Holland, the quantity used is from 14 to 22 lb. (!) per cubic foot. The specifications frequently issued by engineers for sleepers for foreign railways describe them to be entirely of heartwood, and then to be creosoted to the extent of 10 lb. of the oil per cubic foot: this it is impossible to do, the value of the process being in the retention of the sapwood.

It being ascertained a few years since that the centres of some sleepers were not impregnated with the fluid, after the sleeper had been creosoted to the extent of 10 lb. of creosote per cubic foot, Sir Macdonald Stephenson suggested, as a means of obviating that defect, the boring of two holes, 1 inch in diameter, through each sleeper longitudinally, and impregnating up to 12 lb. or 14 lb. per cubic foot. By that means the creosote would be sent all through the sleeper. The boring by hand would be an expensive process, but by machinery it might be effected at a comparatively small increased cost.

During the last twenty-five years an enormous quantity of creosoted railway sleepers have been sent to India and other hot climates. The native woods are generally too hard for penetration. On the great Indian Peninsula Railway the native woods were so hard and close-grained that they could not be impregnated with any preservative substance, sâl wood being principally used, into which creosote would not penetrate more than one quarter of an inch. As regards creosoting wood in India, it is moreover a costly process, owing to the difficulty and expense of conveying creosote from England; iron tanks are necessary to hold the oil when on board ship, and, being unsaleable in India, add to the expense.

English contractors often send piles to be creosoted which have been taken from the timber docks. The large quantity of water they contain resists the entrance of the oil, and the result is that a great deal of timber is badly prepared because the contractors cannot obtain it dry.

In the best creosoting works the tank or cylinder is about

6 feet diameter, and from 20 to 50 feet long. In some instances cylinders are open at both ends, and closed with iron doors, so that sleepers or timber entered at one end on being treated can be delivered finished at the opposite end; but for all practical purposes one open end is sufficient, as the oil when heated being of such a searching character it is a difficult matter to get the doors perfectly air-tight, consequently they are apt to leak during the time the pressure is being applied. Pipes are led from the cylinder to the air and force pumps; the air is not only extracted from the interior of the cylinder, but also from the pores of the timber. When a vacuum is made, the oil, which is contained in a tank below the cylinder, is allowed to rush in, and, as soon as the cylinder is full, the inlet pipe is shut and the pressure pumps started to force the oil into the wood; the pressure maintained is from 150 to 200 lb. to the square inch, until the wood has absorbed the required quantity of oil, which is learned by an index gauge fixed to the working tank below. All cylinders are fitted with safety valves, which allow the oil not immediately absorbed to pass again into the tank. The oil is heated by coils of pipe placed in the tank, through which a current of steam is passed from end to end, raising the temperature to 120°.

With regard to the cost of creosoting: half-round sleepers, being 9 feet long, 10 inches wide, and 5 inches thick, properly creosoted, are worth about 4s. each; adzing for the chairs (done by machine) costs 6s. per 100. These prices, unfortunately, vary very much, according to circumstances. The fir sleepers on the London and Birmingham

Railway cost **7s. 6d.** each, and the patent preservative added 9d. more to the expense, but they did not cost so much on other lines A London builder wrote to us in 1870, as follows: "Our price for creosoting timber, &c., is 15s. per load of 50 cubic feet. Price of creosote, 2d. per gallon."

By returns from the Leith Harbour Works it was shown that the average quantity of creosote absorbed by the timber was $57\frac{7}{8}$ gallons per load, or 577 lb. weight forced into 50 cubic feet of wood. Assuming the cost to be 15s. per load, and the creosote at 2d. per gallon, the creosote would cost 9s. 8d., and the labour and profit 5s. 4d. per load of 50 cubic feet.

It is essential to observe that all methods of protecting timber depend for their success upon the skilful and conscientious manner in which they are applied; for, as they involve chemical actions on a large scale, their efficiency must depend upon the observance of the minute practical precautions required to exclude any disturbing causes. In the case of creosoting: to distil the creosote, to draw the sap or other moisture from the wood, and subsequently to inject the creosote in a proper manner, it is necessary that the operations should be carried into effect under the supervision of experienced persons of high character.

Mr. Bethell's process has been and still is being tested on the Indian railways. According to Dr. Cleghorn, it appears that many of the creosoted sleepers have, however, "been found decayed in the centre, the interior portion being scooped out, leaving nothing but a deceptive shell, in some instances not more than $\frac{1}{2}$ inch in

PATENT PRESERVATIVE SYSTEM.

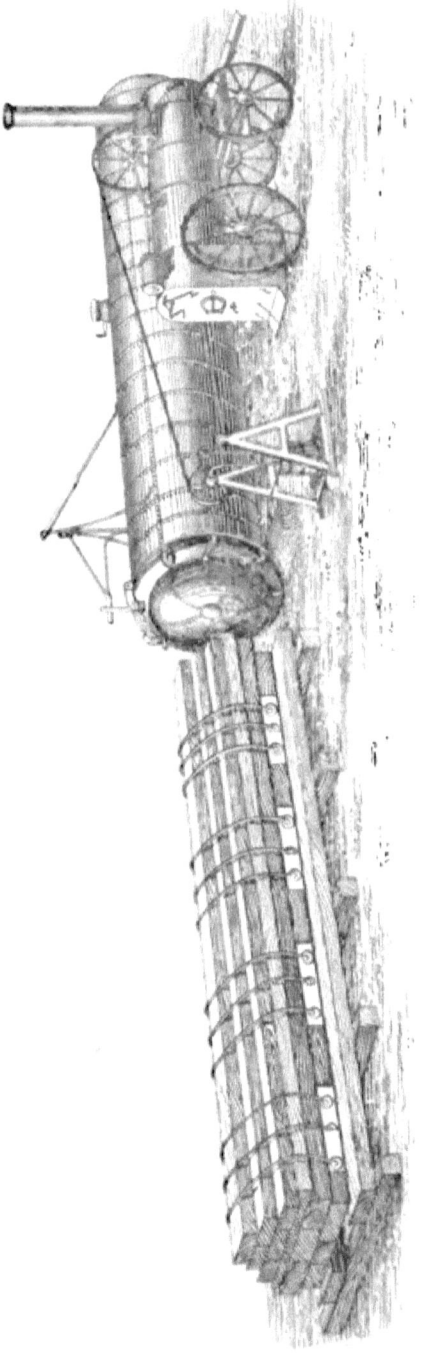

Messrs. John Bethell and Co's. Timber preserving apparatus.

thickness," but he does not state whether the sleepers were prepared in England or in India; because, if prepared in India, it is probable that some of the hard Indian woods, into which it is not possible to get creosote or any other preservative fluid, had been used. Mr. Burt, who has large timber-preserving works in London for creosoting, stated about eight years since, that after an experience of twenty years, during which time he had sent over one million and a half sleepers to India alone, besides having prepared many thousand loads of timber for other purposes, he could safely assert that the instances of failure had been rare and isolated.

A section of a piece of timber impregnated with creosote presents some curious and very distinctive characteristics, according to the duration of the process of injection and amount of tar injected. In every case the injected tar follows the lines and sinuosities of the longitudinal fibres. When injected in sufficient quantity it fills the pores altogether; when, on the contrary, the process has been incompletely performed, which, however, is generally sufficient, the tar accumulates in the transverse sections, and plugs the channels that give access to deleterious agents.

The experiments made by M. Melseuns on oaken blocks exposed to the fumes of *liquid ammonia* show that the conservating fluids follow the precise course that would be taken by decay. In wood treated with creosote the tar acts on the very parts first exposed to injury, and on the course that would be taken by decay, which is thus rendered impossible. The methods of injection suggested by M. Melseuns in 1845 did not answer equally well with every kind of wood. After trying wooden blocks in every sort of condition, dressed and in the rough, green and dry, sound and decayed, M. Melseuns found that alder, birch, beech, hornbeam, and willow were easily and completely impregnated; deal sometimes resisted the process, the innermost layers remaining white; poplar and oak offered a very great resistance — indeed, with poplar it was found necessary to repeat the process.

The decay of sleepers, prepared and unprepared, will often depend on their form. Three forms have been used: 1st, the half-round sleeper, 10 inches by 5 inches; these are now almost universally used; 2nd, the triangular sleeper, about 12 inches wide on each side, used by Mr. Cubitt on the Dover line, but since abandoned; and 3rd, the half square, 14 inches by 7 inches, used by Mr. Brunel and still in use. Mr. G. O. Maun, in reporting on the state of the sleepers of the Pernambuco Railway, states that fair average samples taken out on the 1st December, 1863 (laid in 1857), show that the half-round intermediate sleeper is in the most perfect state of preservation; in fact, nearly as good as on the day it was put down; while the square-sawn or joint sleeper has not withstood the effects of the climate so well.

The kind of ballast in which it will be most advisable to lay the sleeper is another important point to be attended to. About 12 miles of the Pernambuco Railway are entirely laid with creosoted sleepers, principally in white sand. In this description of ballast the half-round sleepers have suffered, since the opening of the first section of the line in 1858 up to 1866, a depreciation of not more than 1 per cent., whilst the square-sawn sleepers have experienced a depreciation of not less than 50 per cent. Had the latter been placed in wet cuttings with ballast retentive of moisture, no doubt the whole of them would have required to be renewed. Hence it is evident that fine open sand ballast, which allows a free drainage during the rains, is best adapted for the preservation of sleepers in the tropics: it has also been found to be the best in most countries.

The number of testimonials given in favour of creosote is very large, and are from the most eminent engineers of all countries, in addition to which Mr. Bethell has received several medals at international exhibitions. The English engineers include Messrs. Brunel, Gregory, Abernethy, Ure, Hemans, Hawkshaw, and Cudworth; the French, MM. Molinos and Forestier; the Dutch, Messrs. Waldorp, Freem, and Von Baumhauer; and the Belgian, M. Crepin. The late Mr. Brunel expressly stated that, in his opinion, well creosoted timbers would be found in a sound and serviceable condition at the expiration of forty years. M. Forestier, French engineer of La Vendée department, reporting to the juries of the French Exhibition of 1867, cites a number of experiments he has lately tried upon many pieces of creosoted and uncreosoted oak, elm, ash, Swedish, Norwegian, and Dantzic red fir, Norway white fir, plane, and poplar, and shows that in each case, except that of the poplar, the resistance of the wood both to bending and crushing weight was much increased by creosoting.

Drs. Brande, Ure, and Letheby, also bear testimony to the efficacy of this mode of preserving timber.

Creosoting has been extensively employed upon all the principal railways in Great Britain. In England, upon the London and North Western, North Eastern, South Eastern, Great Western, &c. In Scotland, on the Caledonian, Great Northern, &c. In Ireland, on the Great Southern and Western, Midland, &c. It has also been and is being employed in Belgium, Holland, France, Prussia, India, and America.

Between the years 1838 and 1840, Sir William Burnett's (formerly Director-General of the Medical Department of the Navy) process was first made known to the public.

This process consists of an injection of *chloride of zinc* into timber, in the proportion of about 1 lb. of the salt to about 9 or 10 gallons of water, forced into the wood under a pressure of 150 lb. per square inch.

The late Professor Graham thus wrote of its efficiency: "After making several experiments on wood prepared by the solution of chloride of zinc for the purpose of preservation, and having given the subject my best consideration, I have come to the following conclusions:

"The wood appears to be fully and deeply penetrated by the metallic salt. I have found it in the centre of a large prepared block.

"The salt, although very soluble, does not leave the wood easily when exposed to the weather, or buried in dry or damp earth. It does not come to the surface of the wood like the crystallizable salts. I have no doubt, indeed, that the greater part of the salts will remain in the wood for years, when employed for railway sleepers or such purposes. This may be of material consequence when the wood is exposed to the attacks of insects, such as the white ant in India, which, I believe, would be repelled by the poisonous metallic salt. After being long macerated in cold water, or even boiled in water, thin chips of the prepared wood retain a sensible quantity of the oxide of zinc; which I confirmed by Mr. Toplis' test, and observed that the wood can be permanently dyed from being charged with a metallic mordant.

"I have no doubt, from repeated observations made during several years, of the valuable preservative qualities of the solution of chloride of zinc, as applied in Sir W. Burnett's process; and would refer its beneficial action chiefly to the small quantity of the metallic salt, which is permanently retained by the ligneous fibre in all circumstances of exposure. The oxide of zinc appears to alter and harden the fibre of the wood, and destroy the solubility, and prevent the tendency to decomposition of the azotised principles it contains by entering into chemical combination with them."

The Report of the Jury, which was drawn up by the Count of Westphalia, at the Cologne International Agricultural Exhibition, in 1865, upon prepared specimens of timber, has the following remarks on the chloride of zinc process:

1st. That chloride of zinc is the only substance which thoroughly penetrates the timber, and is at the same time the best adapted for its preservation.

2nd. That the process of impregnating the wood after cutting is more useful and rational than doing so while the tree is growing.

3rd. That red beech is the only wood which has been impregnated in an uniform and thorough manner.

It should, however, be stated that the Jury had very slender evidence presented to it respecting the creosoting process. The creosoted specimens had been impregnated under the pressure of 60 lb. to 65 lb. per square inch for three or four hours, and were consequently inefficiently done; in England the pressure per square inch would have been at least 140 lb.

Drs. Brande and Cooper, of England, and Dr. Cleghorn, of India, also wrote favourably of Sir W. Burnett's process.

In 1847 a powerful cylinder, of Burnett's construction, hermetically closed, was laid down adjoining the sawmills in Woolwich dockyard. It was found to admit the largest description of timber for the purpose of having the moisture extracted, and the pores filled with chloride of zinc. Three specimens of wood—English oak, English elm, and Dantzic fir—remained uninjured in the fungus pit at Woolwich for five years; while similar, but unprepared, specimens were all found more or less decayed.

The cost of preparing timber by this process is 12s. per load, besides 2s. for landing and loading: 1 lb. of the material costing 1s., which is sufficient for 9 or 10 gallons of water.

Sir W. Burnett and Co.'s works for hydraulic apparatus and tanks are at Nelson Wharf, Millwall, Poplar; their office is at 90, Cannon Street, London. Their terms are—

"For timber, round or square, including planks, deals, **hop-poles**, paving-blocks, &c., **against** rot, 12s. per load of 50 cubic feet.

"For park palings, cabinet work, wine and other **laths, as** per agreement.

"For railway sleepers, 9 feet long, 10 inches by 5 inches, landing and reshipping included, 7d. each.

"For timber to be rendered uninflammable, 25s. per load."

Sir W. Burnett's firm now sell their patent concentrated solution at 5s. per gallon: each gallon must be diluted with 40 gallons of water, according to the instructions in the licence, for which no charge is made.

The reader will probably have observed that this process is *considered* to render timber uninflammable; then let us see what will be the cost of obtaining a fire-proof house.

The principal building material which causes the destruction of our houses by fire is wood—*combustible wood*. If, therefore, (as nearly all our houses are "brick and timber" erections,) we render this wood uninflammable, what will the cost be?

The following is an *approximate* estimate of the extra expense, including sundries, &c.:—

Timber and Deals. Loads.	Cost of House. £	Additional expense. £
25	1000	34
15	600	21
10	400	14
8	250	12

When will the Building Act compel us to use this table *in daily practice?*

Although among the many attempts to preserve wood those in England have proved the most successful, it should be mentioned that France, Germany, and America have given much attention to the subject.

At the end of the last century Du Hamel and Buffon pointed out the possibility of preserving wood, as well as the means of rendering it unalterable. As early as 1758 Du Hamel made experiments on the vital suction of plants, and made some curious observations on the different rings of vegetable matter which absorb most liquid in different plants. He also tried the effect of vital suction and pressure (of gravitation) acting at the same time. His process was reviewed by Barral in 1842.

About 1784 M. Migneron invented a process about which little is now known, but the wood was covered with certain fatty substances. Wood nine years exposed to deterioration was improved by this process. M. Migneron had the approval of Buffon, Franklin, and the Academies. His invention was again brought into notice in 1807, when it was found that timber which had been prepared by it in 1784, and exposed more than twenty years, was quite sound.

In 1811 Cadet de Gassicourt made different kinds of wood imbibe vegetable and mineral substances, and certain unguents: he used metallic salts (iron, tin, &c.).

In 1813 M. Champy plunged wood into a bath of *tallow* at 334°, and kept it there two or three hours. His experiments were afterwards repeated by Mr. Payne.

About the year 1832 it was proposed in America to apply *pyroligneous acid* to the surface of wood, or introduce it by fumigation.

Biot (who has written an excellent life of Sir Isaac Newton) remarked, in 1834, that wood could be soaked by pressure; but his process of penetrating it with liquids was imperfect, and his discovery remains unapplied.

A Frenchman, of the name of Bréant, made about this time a discovery which **preceded** Boucherie's **method,** which is adopted to a great extent in **France.** Bréant's apparatus consisted of a very ingenious machine, which, acting by pressure, caused liquids to penetrate to all points of a mass of **wood of** great diameter and considerable length. He may therefore **be regarded as** having solved the problem of penetration in a scientific, though not in a practically applied, point of view. Dr. **Boucherie** testified before the Académie des Sciences, in 1840, **to the** merit of Bréant's invention, which, with modifications by Payne, Brochard, and Gemini, has been worked in France and England. **This process** was recommended **by Payne** in 1840 and **1844,** and imitated **by** him in France, and later **on by Vengat and Bauner,** who used both *an air pump and a forcing pump*. Bréant obtained three patents, viz. **1st,** in **1831, to act by pressure;** 2nd, in 1837, **by** vital suction; and 3rd, in 1838, vacuum **by** steam. A mixture of linseed oil and resin succeeded best with him. **He attached** more importance to the thorough penetration **of the wood than to the** choice of the penetrating substances. He borrowed his process from Du Hamel, but to make the necessary suction in the pores he produces a partial vacuum in **the** impregnating cylinder by filling **it** with steam, and condensing the steam.

Previous **to** Boucherie's method, **a German,** Frantz Moll, in **1835,** proposed to introduce into wood *creosote in a state of vapour*, but the process was found to be too expensive. This was a modification of Maconochie and Lukin's trials in 1805 and 1811.* A similar process has

* See 'Repertory of **Patent Inventions,**' December, 1836.

since arisen in New York: we believe Mr. Renwick, of that place, suggested it.

Such were the known labours when Dr. Boucherie, in December, 1837, devoted his time to a series of experiments upon timber, with a view to discover some preservative process which should answer the following requirements: First, for protecting wood from dry rot or wet rot; second, for increasing its hardness; third, for preserving and developing its flexibility and elasticity; fourth, for preventing its decay, and the fissures that result from it, when, after having been used in construction, it is left exposed to the variations of the atmosphere; fifth, for giving it various and enduring colours and odours; and sixth and last, for greatly reducing its inflammability.

It is a curious coincidence that at Bordeaux, in 1733, the Academy received a memoir relative to the circulation of the sap and coloured liquids in plants; and it was at Bordeaux, a century afterwards, viz. 1837, that M. Boucherie first mentioned his method.

M. Boucherie's process was first discussed in Paris in June, 1840. It consists in causing a solution of *sulphate of copper* to penetrate to the interior of freshly cut woods, to preserve them from decay; he occasionally used the chloride of calcium, the *pyrolignite of iron (pyrolignite brut de fer), prussiate of iron, prussiate of copper*, and various other metallic salts. As a general rule sulphate of copper is used; but when the hardness of the wood is desired to be increased, pyrolignite of iron is taken (1 gallon of iron to 6 gallons of water); and when the object is to render the wood flexible, elastic, and at the same time uninflammable, chloride of calcium is used. The liquid is taken up by the tree either whilst growing in the earth or immediately after it has been felled. Not more than two or three months should be allowed to elapse before the timber is operated upon, but the sooner it undergoes the process after being felled the better.

Sulphate of copper is said to be superior to corrosive sublimate. Dr. Boucherie's process of the injection of wood with the salts of copper is as simple as it is easy. For those woods intended for poles it consists in plunging the base of a branch, furnished with leaves, into a tub containing the solution. The liquid ascends into the branches by the action of the leaves, and the wood is impregnated with the preservative salt. As for logs, the operation consists in cutting down the tree to be operated upon; fixing at its base a plank, which is fixed by means of a screw placed in the centre, and which can be tightened at will when placed in the centre of the tree. This plank has, on the side to be applied to the bottom of the tree, a rather thick shield of leather, cloth, pasteboard, or some other substance, intended to establish a space between it and the wood, sufficient for the preserving fluid to keep in contact with the freshly cut surface of the tree. The liquid is brought there from a tub or other reservoir, by the help of a slanting pole made on the upper surface of the tree, and in which is put a tube, adapted at its other extremity to a spigot in the upper reservoir which contains the solution. A pressure of 5 mètres suffices; so that the instant the sap of the tree is drawn away it escapes, and is replaced by the liquid saturated with sulphate of copper. The proportion of sulphate of copper in the solution should be 1 lb. of the salt to $12\frac{1}{2}$ gallons of water. As soon as the operation terminates (and it lasts for some hours for the most difficult logs), the wood is ready for use.

For various practical reasons, the first invention of impregnating the wood of the tree whilst still in a growing state, causing it to suck up various solutions by means of the absorbing power of the leaves themselves, was subsequently abandoned; and at the present time a cheap, simple, and effective process is adopted for impregnating the felled timbers with the preserving liquid, designated in France "trait de scie, et la cuisse foulante." The trunk of a newly felled tree is cut into a length suitable for two railway sleepers; a cross cut is made on the prostrate timber to nearly nine-tenths of its diameter; a wedge is then inserted, and a cord is wound round on the cut surface, leaving a shallow chamber in the centre, when it is then closed by withdrawing the wedge. A tube is then inserted through an auger hole into this chamber, and to this tube is attached an elastic connecting tube from a reservoir placed some 20 or 30 feet above the level on which the wood lies, and a stream of the saturating fluid with this pressure passes into the chamber, presses on the sap in the sap tubes, expels it at each end of the tree, and itself supplies its place. The fluid used is a solution of copper in water, in the proportion of 10 or 12 per cent., and a chemical test that ascertains the pressure of the copper solution is applied at each end of the tree from which the sap exudes, by which the operator ascertains when the process is completed.

A full account of this process may be found in the number for June, 1840, of 'Les Annales de Chimie et de Physique.' Messrs. de Mirbel, Arago, Poucelet, Andouin, Gambey, Boussingault, and Dumas, on the part of l'Académie des Sciences, made a report upon Dr. Boucherie's process, confirming the value of the invention. In France, Dr. Boucherie, some years since, relinquished his brevet, and threw the process open to the public, in consideration of a national reward; whilst in England he has obtained two patents (1838 and 1841), which, however, are similar to Bethell's patent, obtained by him on July 11, 1838: *which is the same day and year of Boucherie's patent.* A prize medal was awarded for Dr. Boucherie's process at the Great Exhibition in London, in 1851, and a grande médaille d'honneur, at the Paris Exhibition of 1855. Many thousands of railway sleepers have been prepared by this process, and laid down on the Great Northern Railway of France, and are at present perfectly sound, whilst others not prepared, on the same line, have rotted. Boucherie's process was used on Belgium railways up to 1859; and it is to be regretted that the reasons which led to its abandonment have not been given in the reports of the railway administration, as such reasons would have afforded reliable data for future experimentalists to go upon.

Messrs. Légé and Fleury-Pironnet's patent for the injection of sulphate of copper into beech and poplar is as follows: After the wood is placed, and the opening hermetically sealed, a jet of steam is introduced, intended at first to enter the timber and open its pores for the purpose of obtaining a sudden vacuum, so as to establish at any time a communication between the interior of the cylinder and the cold water condenser; at the same time the air pump is put in action. The vacuum caused is very powerful, and is equal to $25\frac{1}{2}$ ins. of the barometer. Under the double influence of the heat and the vacuum the sap is quickly evaporated from the wood as steam, and ejected from the cylinder by the air pump, so that in a very short time the wood is fully prepared to admit the preserving liquid through the entire bulk.

The use of sulphate of copper for preserving timber has not been, however, confined to France, for about the time Dr. Boucherie brought forward his process, a Mr. Margary took out a patent in England for the use of the same material. His method consists in steeping the substances to be preserved in a solution of sulphate of copper, of the strength of 1 lb. of the sulphate to 8 gallons of water, and leaving them in it till thoroughly saturated. The timber is allowed to remain in the tank two days for every inch of its thickness. Another method is to place the timber in a closed iron vessel of great strength, and it is made to imbibe the solution by exhaustion and pressure, the operation occupying but a short time.

Sulphate of copper is sold in quantities at 4d. per lb.; so that 100l. would buy 6000 lb., and each pound weight is sufficient for 7 or 8 gallons of water, according to Margary; or 12 gallons of water, according to Boucherie.

To preserve railway sleepers, the French railway engineers require $\frac{1}{4}$ lb. of sulphate of copper per cubic foot, say at least 12 lbs. to the load of 50 feet, to be used in a 2 per cent. solution; so that a load of timber can be rendered imperishable for the sum of four shillings, exclusive of labour, if sulphate of copper be reckoned at 4*d.* per lb.

With respect to the use of pyrolignite of iron, Mr. Bethell considers it an expensive process, the pyrolignite costing 6*d.* to 9*d.* per gallon, whilst the oil of tar can be delivered at from 2*d.* to 3*d.* per gallon: the cost of these materials is constantly varying.

A great many sleepers were prepared on the Great Western Railway by pyrolignite of iron, and all have *decayed.* Their black colour makes them exactly resemble creosoted sleepers, and *many mistakes* have arisen from this resemblance.

Messrs. Dorsett and Blythé's (of Bordeaux) patent process of preparing wood by the injection of heated solutions of sulphate of copper is said to have been adopted by French, Spanish, and Italian, as well as other continental railway companies, by the French Government for their navy and other constructions, and by telegraph companies for poles on continental lines. It is as cheap as creosote, and is employed in places where creosote cannot be had. Wood prepared by it is rendered incombustible. Wood for outdoor purposes so prepared has a clean yellowish surface, without odour; it requires no painting, remains unchangeable for any length of time, and can be employed for any purpose, the same as unprepared material, and carried with other cargo without hindrance.* Messrs.

* See 'Étuves de Désiccation et Appareil pour l'Injection des Bois.' Par MM. Dorsett et Blythé, manufacturiers, à Bordeaux. 1859.

Dorsett and Blythé's process is similar to that of Mr. Knab, which consisted of a solution of sulphate of copper, heated to nearly boiling point, and placed in a lead cylinder, protected by wood.

In 1846, 80,000 sleepers, treated with sulphate of copper, were laid down on French railways, and after nine years' exposure were found to be as perfect as when first laid.

Mr. H. W. Lewis, University of Michigan, U.S., thus writes in the 'Journal' of the Franklin Institute, in 1866, with reference to the decay of American railway sleepers: "Allowing 2112 sleepers per mile, at 50 cents each, 1056 dols. per mile of American railroad decay every seven years. Thoroughly impregnate those sleepers with sulphate of copper, at a cost of 5 cents each, and they would last twice as long. Thus would be effected a saving of 880 dols. per mile in the seven years on sleepers alone. In the United States, there are 33,906·6 miles of railroad. The whole saving on these lines would be 29,389,568 dols., or upwards of 4,262,795 dols. per annum."

With reference to the decay of unprepared wooden sleepers, it may be here stated that the renewal of wooden sleepers on the Calcutta and Delhi Indian line alone costs annually 130,000*l*.

The preservative action of sulphate of copper on wood has long been known, but there are several things in its action which require explanation. The 'London Review' says that Kœnig has lately investigated the chemical reactions which occur when wood is impregnated with a preservative solution of blue vitriol. He finds, as a general rule, that a certain quantity of basic sulphate of copper remains combined in the pores of the wood in such a manner that it cannot be washed out with water. The copper salt may be seen by its green colour in the spaces between the yearly rings in the less compact portions of the wood, that is to say, in those portions which contain the sap. Those varieties of wood which contain the most resin retain the largest amount of the copper salt—oak, for example, retaining but little of it. The ligneous fibre itself appears to have little or nothing to do with the fluxation of the copper salt, and indeed none whatever is retained in chemical combination, so that it cannot be washed out with water, by pure cellulose. When wood, from which all resin has been extracted by boiling alcohol, is impregnated with sulphate of copper, it does not become coloured like the original resinous wood, and the copper salt contained in it may be readily washed out with water. In like manner, from impregnated resinous wood all the copper salt may be removed, with the resin, by means of alcohol. The constituents of the blue vitriol are consequently fixed in the wood by means of the resin which this contains. Further, it is found that the impregnated wood contains less nitrogen than that which is unimpregnated, and that it is even possible to remove all the nitrogenous components of the wood by long-continued treatment with the solution of sulphate of copper; the nitrogenous matters being soluble in an excess of this solution, just as the precipitate which forms when aqueous solutions of albumen and sulphate of copper are mixed is soluble in excess of the latter. Since the nitrogenous matters are well known to be promoters of putrefaction, their removal readily accounts for the increased durability of the impregnated wood. The utility of blue vitriol as a preservative may also depend on a measure upon the resinous copper salt which is formed, by which the pores of the wood are more or less filled up, and the ligneous fibre covered, so that contact with the air is prevented, and the attack of insects hindered. It is suggested that those cases in which the anticipated benefits have not been realized in practice, by impregnating wood with a solution of blue vitriol, may probably be referred to the use of an iusufficient amount of this agent; that is, where the wood was not immersed in the solution for a sufficient length of time. The action should be one of lixiviation, not merely of absorption.

In 1841, a German, named Müenzing, a chemist of Heibronn, proposed *chloride of manganese* (waste liquor in the manufacture of bleaching powder) as a preservative against dry rot in timber; but his process has not been adopted in England, and very little noticed abroad.

In July, 1841, Mr. Payne patented his invention for *sulphate of iron* in London; and in June and November, 1846, in France; and in 1846 in London, for *carbonate of soda*.* The materials employed in Payne's process are sulphate of iron and sulphate of lime, both being held in solution with water. The timber is placed in a cylinder in which a vacuum is formed by the condensation of steam, assisted by air pumps; a solution of sulphate of iron is then admitted into the vessel, which instantly

* See 'Repertory of Patent Inventions,' April, 1847.

insinuates itself into all the pores of the wood, previously freed from air by the vacuum, and, after about a minute's exposure, impregnates its entire substance; the *sulphate of iron* is then withdrawn, and another solution of *sulphate of lime* thrown in, which enters the substance of the wood in the same manner as the former solution, and the two salts react upon each other, and form two new combinations within the substance of the wood—muriate of iron, and muriate of lime. One of the most valuable properties of timber thus prepared is its perfect incombustibility: when exposed to the action of flame or strong heat, it simply smoulders, and emits no flame. We may also reasonably infer that with such a compound in its pores, decay must be greatly retarded, and the liability to worms lessened, if not prevented. The greatest drawback consists in the increased difficulty of working. This invention has been approved by the Commissioners of Woods and Forests, and has received much approbation from the architectural profession. Mr. Hawkshaw, C.E., considers that this process renders wood brittle. It was employed for rendering wood uninflammable in the Houses of Parliament (we presume, in the carcase; for *steaming* was used for the joiner's work), British Museum, and other public buildings; and also for the Royal Stables at Claremont.

In 1842, Mr. Bethell stated before the Institute of Civil Engineers, London, that *silicate of potash*, or *soluble glass*, rendered wood uninflammable.

In 1842, Professor Brande proposed *corrosive sublimate* in *turpentine*, or *oil of tar*, as a preservative solution.

In 1845, Mr. Ransome suggested the application of *silicate of soda*, to be afterwards decomposed by an acid in the fibre of the wood; and in 1846, Mr Payne proposed soluble sulphides of the earth (*barium sulphide*, &c.), to be also afterwards decomposed in the woods by acids.

In 1855, a writer in the 'Builder' suggested an equal mixture of alum and borax (biborate of soda) to be used for making wood uninflammable. We have no objection to the use of alum and borax to render wood uninflammable, providing it does not *hurt the wood*.

Such are the *principal* patents, suggestions, and inventions, up to the year 1856; but there are many more which have been brought before the public, some of which we will now describe.

Dr. Darwin, some years since, proposed absorption, first, of *lime water*, then of a weak solution of *sulphuric acid*, drying between the two, so as to form a gypsum (sulphate of lime) in the pores of the wood, the latter to be previously well seasoned, and when prepared to be used in a dry situation.

Dr. Parry has recommended a preparation composed of *bees-wax*, *roll brimstone*, and *oil*, in the proportion of 1, 2, and 3 ounces to $\frac{3}{4}$ gallon of water; to be boiled together and laid on hot.

Mr. Pritchard, C.E., of Shoreham, succeeded in establishing *pyrolignite of iron* and *oil of tar* as a preventive of dry rot; the pyrolignite to be used very pure, the oil applied afterwards, and to be perfectly free from any particle of ammonia.

Mr. Toplis recommends the introduction into the pores of the timber of a solution of sulphate or muriate of iron; the solution may be in the proportion of about 2 lb. of the salt to 4 or 5 gallons of water.

An invention has been lately patented by Mr. John Cullen, of the North London Railway, Bow, for preserving wood from decay. The inventor proposes to use a composition of *coal-tar, lime,* and *charcoal;* the charcoal to be reduced to a fine powder, and also the lime. These materials to be well mixed, and subjected to heat, and the wood immersed therein. The impregnation of the wood with the composition may be materially aided by means of exhaustion and pressure. Wood thus prepared is considered to be proof against the attacks of the white ant.

The process of preserving wood from decay invented by Mr. L. S. Robins, of New York, was proposed to be worked extensively by the "British Patent Wood Preserving Company." It consists in first removing the surface moisture, and then charging and saturating the wood with hot *oleaginous vapours* and compounds. As the Robins' process applies the preserving material in the form of vapour, the wood is left clean, and after a few hours' exposure to the air it is said to be fit to be handled for any purposes in which elegant workmanship is required. Neither science nor extraordinary skill is required in conducting the process, and the treatment under the patent is said to involve only a trifling expense.

Reference has already been made to the use of *petroleum*. The almost unlimited supply of it within the last few years has opened out a new and almost boundless source of wealth. An invention has been patented in the name of Mr. A. Prince, which purports to be an improvement in the mode of preserving timber by the aid of petroleum. The invention consists, firstly, in the immersion of the timber in a suitable vessel or receptacle, and to exhaust the air therefrom, by the ordinary means of preserving wood by saturation. The crude petroleum is next conveyed into the vessel, and thereby caused to penetrate into every pore or interstice of the woody fibre, the effect being, it is said, to thoroughly preserve the wood from decay. He also proposes to mix any cheap mineral paint or pigment with crude petroleum to be used as a coating for the bottom of ships before the application of the sheathing, and also to all timber for building or other purposes. The composition is considered to render the timber indestructible, and to repel the attacks of insects. Without expressing any opinion upon this patent as applied to wood for building purposes, we must again draw attention to the high inflammability of petroleum.

The 'Journal' of the Board of Arts and Manufactures for Upper Canada considers the following to be the cheapest and the best mode of preserving timber in Canada: Let the timbers be placed in a drying chamber for a few hours, where they would be exposed to a temperature of about 200°, so as to drive out all moisture, and by heat, coagulate the albuminous substance, which is so productive of decay. Immediately upon being taken out of the drying chamber, they should be thrown into a tank containing crude petroleum. As the wood cools, the air in the pores will contract, and the petroleum occupy the place it filled.

Such is the extraordinary attraction shown by this substance for dry surfaces, that by the process called capillary attraction, it would gradually find its way into the interior of the largest pieces of timber, and effectually coat the walls and cells, and interstitial spaces. During the lapse of time, the petroleum would absorb oxygen, and become inspissated, and finally converted into a bituminous substance, which would effectually shield the wood from destruction by the ordinary processes of decay. The process commends itself on account of its cheapness. A drying chamber might easily be constructed of sheet iron properly strengthened, and petroleum is very abundant and accessible. Immediately after the pieces of timber have been taken out of the petroleum vat, they should be sprinkled with wood ashes in order that a coating of this substance may adhere to the surface, and carbonate of potash be absorbed to a small depth. The object of this is to render the surface incombustible; and dusting with wood ashes until quite dry will destroy this property to a certain extent.

The woodwork of farm buildings in this country is sometimes subjected to the following: Take two parts of *gas-tar*, one part of *pitch*, one part *half caustic lime* and *half common resin*; mix and boil these well together, and put them on the wood quite hot. Apply two or three coats, and while the last coat is still warm, dash on it a quantity of well-washed sharp sand, previously prepared by being sifted through a sieve. The surface of the wood will then have a complete stone appearance, and may be durable. It is, of course, necessary, that the wood be perfectly dry, and one coat should be well hardened before the next is put on. It is necessary, by the use of lime and long boiling, to get quit of the ammonia of the tar, as it is considered to injure the wood.

Mr. Abel, the eminent chemist to the War Department, recommends the application of *silicate of soda* in solution, for giving to wood, when applied to it like paint, a hard coating, which is durable for several years, and is also a considerable protection against fire. The silicate of soda, which is prepared for use in the form of a thick syrup, is diluted in water in the proportion of 1 part by measure of the syrup to 4 parts of water, which is added slowly, until a perfect mixture is obtained by constant stirring. The wood is then washed over *two* or *three* times with this liquid by means of an ordinary whitewash brush, so as to absorb as much of it as possible. When this first coating is nearly dry, the wood is painted over with *another* wash made by slaking good fat lime, diluted to the consistency of thick cream. Then, after the limewash has become moderately dry, *another* solution of the silicate of soda, in the proportion of 1 of soda to 2 of water, is applied in the same manner as the first coating. The preparation of the wood is then complete; but if the lime coating has been applied too quickly, the surface of the wood may be found, when quite dry, after the last coating of the silicate, to give off a little lime when rubbed with the hand; in which case it should be *once more* coated over with a solution of the silicate of the same strength as in the first operation. If Mr. Abel had been an architect or builder, he would never have invented this process. What would the cost be? and would not a special clerk of the works be necessary to carry out this method in practice?

The following coating for piles and posts, to prevent them from rotting, has been recommended on account of its being economical, impermeable to water, and nearly as hard as stone: Take 50 parts of *resin*, 40 of *finely powdered chalk*, 300 parts of *fine white sharp sand*, 4 parts of *linseed oil*, 1 part of native *red oxide of copper*, and 1 part of *sulphuric acid*. First, heat the resin, chalk, sand, and oil, in an iron boiler; then add the oxide, and, with care, the acid; stir the composition carefully, and apply the coat while it is still hot. If it be not liquid enough, add a little more oil. This coating, when it is cold and dry, forms a varnish which is as hard as stone.

Another method for fencing, gate-posts, garden stakes, and timber which is to be buried in the earth, may be mentioned. Take **11 lb. of** *blue vitriol* (sulphate of copper) and 20 quarts of water; dissolve the vitriol with boiling water, and then add the remainder of the water. The end of the wood is then to be put into the solution, and left to stand four or five days; for shingle, three days will answer, and for posts, 6 inches square, ten days. Care should be taken that the saturation takes place in a well-pitched tank or keyed box, for the reason that any barrel will be shrunk by the operation so as to leak. Instead of expanding an old cask, as other liquids do, this shrinks it. This solution has also been used in dry rot cases, when the wood is only slightly affected.

It will sometimes be found that when oak fencing is put up new, and tarred or painted, a fungus will vegetate through the dressing, and the interior of the wood be rapidly destroyed; but when undressed it seems that the weather desiccates the gum or sap, and leaves only the woody fibre, and the fence lasts for many years.

About fifteen years ago, Professor Crace Calvert, F.R.S., made an investigation for the Admiralty, of the qualities of different woods used in ship-building. He found the goodness of teak to consist in the fact that it is highly charged with *caoutchouc*; and he considered that if the tannin be soaked out of a block of oak, it may then be interpenetrated by a *solution of caoutchouc*, and thereby rendered as lasting as teak.

We can only spare the space for a few words about this method.

1st. We have seen lead which has formed part of the gutter of a building previous to its being burnt down: lead melts at 612° F.; caoutchouc at 248° F.; therefore caoutchouc would not prevent wood from being destroyed by fire. At 248° caoutchouc is highly inflammable, burns with a white flame and much smoke.

2nd. We are informed by a surgical bandage-maker of high repute, that caoutchouc, when used in elastic knee-caps, &c., *will perish*, if the articles are left in a drawer for two or three years. When hard, caoutchouc is brittle.

Would it be advisable to interpenetrate oak with a solution of caoutchouc? In 1825, Mr. Hancock proposed a solution of 1½ lb. of caoutchouc in 3 lb. of essential oil, to which was to be added 9 lb. of tar. Mr. Parkes, in 1843, and M. Passez, in 1845, proposed to dissolve caoutchouc in sulphur: painting or immersing the wood. Maconochie, in 1805, after his return from India, proposed distilled *teak* chips to be injected into fir woods.

Although England has been active in endeavouring to discover the best and cheapest remedy for dry rot, France has also been active in the same direction.

M. le Comte de Chassloup Lambat, Member of the late Imperial Senate of France, considers that, as *sulphur* is most prejudicial to all species of fungi, there might, perhaps, be some means of making it serviceable in the preservation of timber. We know with what success it is used in medicine. It is also known that coopers burn a sulphur match in old casks before using them—a practice which has evidently for its object the prevention of mustiness, often microscopic, which would impart a bad flavour to the wine.

M. de Lapparent, late Inspector-General of Timber for the French Navy, proposed to prevent the growth of fungi by the use of a paint having flour of sulphur as a basis, and linseed oil as an amalgamater. In 1862 he proposed charring wood; we have referred to this process in our last chapter (p. 96).

The paint was to be composed of:

Flour of sulphur	200 grammes	3,088 grains.
Common linseed oil	135 ,,	2,084 ,,
Prepared oil of manganese	30 ,,	463 ,,

He considered that by smearing here and there either the surfaces of the ribs of a ship, or below the ceiling, with this paint, a slightly sulphurous atmosphere will be developed in the hold, which will purify the air by destroying, at least in part, the sporules of the fungi. He has since stated that his anticipations have been fully realized. M. de Lapparent also proposes to prevent the decay of timber by subjecting it to a skilful carbonization with common inflammable coal gas. An experiment was made at Cherbourg, which was stated to be completely successful. The cost is only about 10 cents per square yard of framing and planking.* M. de Lapparent's gas method is useful for burning off old paint. We saw it in practice (April, 1875) at Waterloo Railway Station, London, and it appeared to be effective.

At the suggestion of MM. Le Châtelier (Engineer-in-chief of mines) and Flachat, C.E.'s, M. Rance, a few years since, injected in a Légé and Fleury cylinder certain pieces of white fir, red fir, and pitch pine with *chloride of sodium*, which had been deprived of the manganesian salts it contained, to destroy its deliquescent property. Some pieces were injected four times, but the greatest amount of solution injected into pitch pine heartwood was from 3 to 4 per cent., and very little more was injected into the white and red fir heartwood. It was also noticed that sapwood, after being injected four times, only gained 8 per cent. in weight in the last three operations. The experiments made to test the relative incombustibility of the injected wood showed that the process was a complete failure; the prepared wood burning as quickly as the unprepared wood.

M. Paschal le Gros, of Paris, has patented his system for

* See Chap. IV., p. 97.

preserving all kinds of wood, by means of a *double salt of manganese* and of *zinc*, used either alone or with an admixture of *creosote*. The solution, obtained in either of the two ways, is poured into a trough, and the immersion of the logs or pieces of wood is effected by placing them vertically in the trough in such a manner that they are steeped in the liquid to about three-quarters of their length. The wood is thus subjected to the action of the solution during a length of time varying from twelve to forty-eight hours. The solution rises in the fibres of the wood, and impregnates them by the capillary force alone, without requiring any mechanical action. The timber is said to become incombustible, hard, and very lasting.

M. Fontenay, C.E., in 1832, proposed to act upon the wood with what he designated *metallic soap*, which could be obtained from the residue in greasing boxes of carriages; also from the acid remains of *oil, suet, iron*, and *brass dust*; all being melted together. In 1816 Chapman tried experiments with *yellow soap;* but to render it sufficiently fluid it required forty times its weight of water, in which the quantity of resinous matter and tallow would scarcely exceed $\frac{1}{80}$th; therefore no greater portion of these substances could be left in the pores of the wood, which could produce little effect.

M. Letellier, in 1837, proposed to use *deuto-chloride of mercury* as a preservative for wood.

M. Dondeine's process was formerly used in France and Germany. It is a paint, consisting of many ingredients, the principal being *linseed oil, resin, white lead, vermilion, lard, and oxide of iron.* All these are to be well mixed, and reduced by boiling to one-tenth, and then applied with a brush. If applied cold, a little varnish or turpentine to be added.

Little is known in England of the inventions which have arisen in foreign countries not already mentioned.

M. Szerelmey, a Hungarian, proposed, in 1868, *potassa, lime, sulphuric acid, petroleum,* &c., to preserve wood.

In Germany, the following method is sometimes used for the preservation of wood: Mix 40 parts of *chalk*, 40 parts of *resin*, 4 of *linseed oil;* melting them together in an iron pot; then add 1 part of native *oxide of copper,* and afterwards, carefully, 1 part of *sulphuric acid.* The mixture is applied while hot to the wood by means of a brush, and it soon becomes very hard.*

Mr. Cobley, of Meerholz, Hesse, has patented the following preparation. A strong solution of *potash, baryta, lime, strontia,* or any of their salts, are forced into the pores of timber in a close iron vessel by a pump. After this operation, the liquid is run off from the timber, and *hydro-fluo-silicic acid* is forced in, which, uniting with the salts in the timber, forms an insoluble compound capable of rendering the wood uninflammable.

About the year 1800, Neils Nystrom, chemist, Norkopping, recommended a solution of *sea salt and copperas,* to be laid upon timber as hot as possible, to prevent rottenness or combustion. He also proposed a solution of *sulphate of iron, potash, alum,* &c., to extinguish fires.

M. Louis Vernet, Buenos Ayres, proposed to preserve timber from fire by the use of the following mixture: Take

---

* See coating for piles, p. 161.

1 lb. of *arsenic*, 6 lb. of *alum*, and 10 lb. of *potash*, in 40 gallons of water, and mix with *oil*, or any suitable tarry matters, and paint the timber with the solution. We have already referred to the conflicting evidence respecting alum and water for wood: we can now state that Chapman's experiments proved that *arsenic* afforded no protection against dry rot. Experiments in Cornwall have proved that where arsenical ores have lain on the ground, vegetation will ensue in two or three years after removal of the ore. If, therefore, alum or arsenic have no good effect on timber with respect to the dry rot, we think the use of both of them together would certainly be objectionable.

The last we intend referring to is a composition frequently used in China, for preserving wood. Many buildings in the capital are painted with it. It is called *Schoieao*, and is made with 3 parts of blood deprived of its febrine, 4 parts of lime and a little alum, and 2 parts of liquid silicate of soda. It is sometimes used in Japan.

It would be practically useless to quote any further remedies, and the reader is recommended to carefully study those quoted in this chapter, and of their utility to judge for himself, bearing in mind those principles which we have referred to before commencing to describe the patent processes. A large number of patents have been taken out in England for the preservation of wood by preservative processes, but only two are now in use,—that is, to any extent,—viz. Bethell's and Burnett's. Messrs. Bethell and Co. now impregnate timber with *copper*, *zinc*, **corrosive** *sublimate, or creosote*; the four best patents.

We insert here a short analysis of different *methods* proposed for seasoning timber :—

Vacuum and Pressure Processes generally.	Vacuum by Condensation of Steam.
Bréant's.	Tissier.
Bethell's.	Bréant.
Payne's.	Payne.
Perin's.	Renard Perin, 1848.
Tissier's.	Brochard and Watteau, 1847.

Separate Condenser.
Tissier.

Employ Sulphate of Copper in closed vessels.	Current of Steam.
Bethell's Patent, 11th July, 1838.	Moll's Patent, 19th January, 1835.
Tissier, 22nd October, 1844.	Tissier's „ 22nd October, 1844.
Molin's Paper, 1853.	Payne's „ 14th Nov., 1846.
Payen's Pamphlet.	Meyer d'Uslaw, 2nd January, 1851.
Légé and Fleury's Pamphlet.	Payen's Pamphlet.

Hot Solution.
Tissier's Patent, 22nd October, 1844.
Knab's Patent, 8th September, 1846.

Most solutions used are heated.

The following are the chief *ingredients* which have been recommended, and some of them tried, to prevent the decomposition of timber, and the growth of fungi :—

Acid, Sulphuric.	Salt, Selenites.
„ Vitriolic.	Oil, Vegetable.
„ of Tar.	„ Animal.
Carbonate of Potash.	„ Mineral.
„ Soda.	Muriate of Soda.
„ Barytes.	Marcosites, Mundic.
Sulphate of Copper.	„ Barytes.
„ Iron.	Nitrate of Potash.
„ Zinc.	Animal Glue.
„ Lime.	„ Wax.
„ Magnesia.	Quick Lime.
„ Barytes.	Resins of different kinds.
„ Alumina.	Sublimate, Corrosive.
„ Soda.	Peat Moss.
Salt, Neutral.	

For the *non-professional* reader we find we have *three* facts:

1st. The most successful patentees have been Bethell and Burnett, in England; and Boucherie, in France: all B's.

2nd. The most successful patents have been *knighted*. Payne's patent was, we believe, used by Sirs R. Smirke and C. Barry; Kyan's, by Sir R. Smirke; Burnett's, by Sirs M. Peto, P. Roney, and H. Dryden; while Bethell's patent can claim Sir I. Brunel, and many other knights. We believe Dr. Boucherie received the Legion of Honour in France.

3rd. There are only at the present time three timber-preserving works in London, and they are owned by Messrs. Bethell and Co., Sir F. Burnett and Co., and Messrs. Burt, Boulton, and Co.: all names commencing with the letter B.

For the *professional* reader we find we have *three hard* facts:

The most successful patents may be placed in three classes, and we give the key-note of their success.

1st. ONE MATERIAL AND ONE APPLICATION.—Creosote, Petroleum. *Order*—Ancient Egyptians, or Bethell's, Burmese.

2nd. TWO MATERIALS AND ONE APPLICATION.—Chloride of zinc and water; sulphate of copper and water; corrosive sublimate and water. *Order*—Burnett, Boucherie, Kyan.

3rd. TWO MATERIALS AND TWO APPLICATIONS.—Sulphate of iron and water; afterwards sulphate of lime and water. Payne.

We thus observe there are *twice three* successful patent processes.

Any inventions which cannot be brought under these three classes have had a short life; at least, we think so.

The same remarks will apply to *external* applications for wood—for instance, coal-tar, *one application,* is more used for fencing than any other material.

We are much in want of a valuable series of experiments on the application of various chemicals on wood to resist *burning to pieces;* without causing it to *rot speedily.*

CHAPTER VI.

ON THE MEANS OF PREVENTING DRY ROT IN MODERN HOUSES; AND THE CAUSES OF THEIR DECAY.

ALTHOUGH writers on **dry rot have generally deemed it a new disease, there is foundation to believe that it pervaded the British Navy in the reign of Charles II.** "Dry rot received a little attention," so writes Sir John Barrow, "about the middle of the last century, at some period of Sir John Pringle's presidency of the Royal Society of London." As timber trees were, no doubt, subject to the same **laws and conditions** 500 **years ago as they are at the present day, it is indeed extremely** probable that **if at** that time unseasoned timber **was used,** and subjected **to heat** and moisture, **dry rot made its** appearance. We propose in **this** chapter **to direct** attention **to the several causes of the** decay of wood, which **by proper building** might be averted.

The necessity of proper ventilation round **the timbers of a building has been** repeatedly advised **in this** volume; for even timber **which has been naturally** seasoned **is at all times** disposed **to resume, from a warm** and stagnant atmosphere, **the elements of decay.** We cannot therefore agree with the following passage from Captain E. M. Shaw's book on **'Fire Surveys,'** which is to be found at page **44 :—"** Circulation of air should on no account be permitted in any part of a building not exposed to view, especially under floors, or inside skirting boards, or wainscots." In the course of this chapter, the evil results from a want of a proper circulation of air will be shown.

In warm cellars, or any close confined situations, where the air is filled with vapour without a current to change it, dry rot proceeds with astonishing rapidity, and the timber work is destroyed in a very short time. The bread rooms of ships; behind the skirtings, and under the wooden floors, or the basement stories of houses, particularly in kitchens, or other rooms where there are constant fires; and, in general, in every place where wood is exposed to warmth and damp air, the dry rot will soon make its appearance.

All kinds of stoves are sure to increase the disease if moisture be present. The effect of heat is also evident from the rapid decay of ships in hot climates; and the warm moisture given out by particular cargoes is also very destructive. Hemp will, without being injuriously heated, emit a moist warm vapour: so will pepper (which will affect teak) and cotton. The ship 'Brothers,' built at Whitby, of green timber, proceeded to St. Petersburgh for a cargo of hemp. The next year it was found on examination that her timbers were rotten, and all the planking, except a thin external skin. It is also an important fact that rats very rarely make their appearance in dry places: under floors they are sometimes very destructive.

As rats will sometimes destroy the structural parts of wood framing, a few words about them may not be out of place. If poisoned wheat, arsenic, &c., be used, the creatures will simply eat the things and die under the floor, causing an intolerable stench. The best method is to make a small hole in a corner of the floor (unless they make it themselves) large enough to permit them to come up; the following course is then recommended:—Take oil of amber and ox-gall in equal parts; add to them oatmeal or flour sufficient to form a paste, which divide into little balls, and lay them in the middle of the infested apartment *at night* time. Surround the balls with a number of saucers filled with water—the smell of the oil is sure to attract the rats, they will greedily devour the balls, and becoming intolerably thirsty will drink till they die on the spot. They can be buried in the morning.

Building timber into new walls is often a cause of decay, as the lime and damp brickwork are active agents in producing putrefaction, particularly where the scrapings of roads are used, instead of sand, for mortar. Hence it is that bond timbers, wall plates, and the ends of girders, joists, and lintels are so frequently found in a state of decay. The ends of brestsummers are sometimes cased in sheet lead, zinc, or firebrick, as being impervious to moisture. The old builders used to bed the ends of girders and joists in loam instead of mortar, as directed in the Act of Parliament, 19 Car. II. c. 3, for rebuilding the City of London.

In Norway, all posts in contact with the earth are carefully wrapped round with flakes of birch bark for a few inches above and below the ground.

Timber that is to lie in mortar—as, for instance, the

ends of joists, door sills and frames of doors and windows, and the ends of girders—if pargeted over with hot pitch, will, it is said, be preserved from the effects of the lime. In taking down, some years since, in France, some portion of the ancient Château of the Roque d'Oudres, it was found that the extremities of the oak girders were perfectly preserved, although these timbers were supposed to have been in their places for upwards of 600 years. The whole of these extremities buried in the walls were completely wrapped round with plates of cork. When demolishing an ancient Benedictine church at Bayonne, it was found that the whole of the fir girders were entirely worm eaten and rotten, with the exception, however, of the bearings, which, as in the case just mentioned, were also completely wrapped round with plates of cork. These facts deserve consideration.

If any of our professional readers should wish to try cork for the ends of girders, they will do well to choose the Spanish cork, which is the best.

In this place it may not be amiss to point out the dangerous consequences of building walls so that their principal support depends on timber. The usual method of putting bond timber into walls is to lay it next the inside; this bond often decays, and, of course, leaves the walls resting only upon the external course or courses of brick; and fractures, bulges, or absolute failures are the natural consequences. This evil is in some degree avoided by placing the bond in the middle of the wall, so that there is brickwork on each side, and by not putting continued bond for nailing the battens to. We object to placing bond in the middle of a wall: the best way, where it can be managed, is to corbel out the wall, resting the ends of the joists on the top course of bricks; thus doing away with the wood-plate. In London, wood bond is prohibited by Act of Parliament, and hoop-iron bond (well tarred and sanded) is now generally used. The following is an instance of the bad effects of placing wood bond in walls: In taking down portions of the audience part and the whole of the corridors of the original main walls of Covent Garden Theatre, London, in 1847, which had only been built about thirty-five years, the wood horizontal bond timbers, although externally appearing in good condition, were found, on a close examination by Mr. Albano, much affected by shrinkage, and the majority of them quite rotten in the centre, consequently the whole of them were ordered to be taken out in short lengths, and the space to be filled in with brickwork and cement.

Some years since we had a great deal to do with "Fire Surveys;" that is to say, surveying buildings to estimate the cost of reinstating them after being destroyed by fire; and we often noticed that the wood bond, being rotten, was seriously charred by the fire, and had to be cut out in short lengths, and brickwork in cement "pinned in" in its place. Brestsummers and story posts are rarely sufficiently burnt to affect the stability of the front wall of a shop building.

In bad foundations, it used to be common, before concrete came into vogue, to lie planks to build upon. Unless these planks were absolutely wet, they were certain to rot in such situations, and the walls settled; and most likely irregularly, rending the building to pieces. Instances of such kind of failure frequently occur. It was found necessary, a few years since, to underpin three of the large houses in Grosvenor Place, London, at an immense expense. In one of these houses the floors were not less than three inches out of level, the planking had been seven inches thick, and most of it was completely rotten: it was of yellow fir. A like accident happened to Norfolk House, St. James's Square, London, where oak planking had been used.

As an example of the danger of trusting to timber in supporting heavy stone or brickwork, the failure of the curb of the brick dome of the church of St. Mark, at Venice, may be cited. This dome was built upon a curb of larch timber, put together in thicknesses, with the joints crossed, and was intended to resist the tendency which a dome has to spread outwards at the base. In 1729, a large crack and several smaller ones were observed in the dome. On examination, the wooden curb was found to be in a completely rotten state, and it was necessary to raise a scaffold from the bottom to secure the dome from ruin. After it was secured from falling, the wooden curb was removed, and a course of stone, with a strong band of iron, was put in its place.

It is said that another and very important source of destruction is the applying end to end of two different kinds of wood: oak to fir, oak to teak or lignum vitæ; the harder of the two will decay at the point of juncture.

The bad effects resulting from damp walls are still further increased by hasty finishing. To enclose with plastering and joiners' work the walls and timbers of a building while they are in a damp state is the most certain means of causing the building to fall into a premature state of decay.

Mr. George Baker, builder of the National Gallery, London, remarked, in 1835, "I have seen the dry rot all over Baltic timber in three years, in consequence of putting it in contact with moist brickwork; the rot was caused by the badness of the mortar, it was so long drying."

Slating the external surface of a wall, to keep out the rain or damp, is sometimes adopted: a high wall (nearly facing the south-west) of a house near the north-west corner of Blackfriars Bridge, London, has been recently slated from top to bottom, to keep out damp.

However well timber may be seasoned, if it be employed in a damp situation, decay is the certain consequence; therefore it is most desirable that the neighbourhood of buildings should be well drained, which would not only prevent rot, but also increase materially the comfort of those who reside in them. The drains should be made water-tight wherever they come near to the walls; as walls, particularly brick walls, draw up moisture to a very considerable height: very strict supervision should be placed over workmen while the drains of a building are being laid. Earth should never be suffered to rest against walls, and the sunk stories of buildings should always be surrounded by an open area, so that the walls may not absorb moisture from the earth: even open areas require to be properly built. We will quote a case to explain our meaning. A house was erected about eighteen months ago, in the south-east part of London, on sloping ground. Excavations were made for the basement floor, and a dry area, "brick thick, in cement," was built at the back and side of the house, the top of the area wall being covered with a stone coping; we do not know whether the bottom of the area was drained. On the top of the coping was placed mould, forming one of the garden beds for flowers. Where the mould rested against the walls, damp entered. The area walls should have been built, in the first instance, above the level of the garden-ground—which has since been done—otherwise, in course of time, the ends of the next floor joists would have become attacked by dry rot.

Some people imagine that if damp is in a wall the best way to get rid of it is to seal it in, by plastering the inside and stuccoing the outside of the wall; this is a great mistake; damp will rise higher and higher, until it finds an outlet; rotting in the meanwhile the wood bond and ends of all the joists. We were asked recently to advise in a curious case of this kind at a house in Croydon. On wet days the wall (*stucco*, outside; *plaster*, inside) was perfectly wet: bands of soft red bricks in wall, at intervals, were the culprits. To prevent moisture rising from the foundations, some substance that will not allow it to pass should be used at a course or two above the footings of the walls, but it should be below the level of the lowest joists. "Taylor's damp course" bricks are good, providing the air-passages in them are kept free for air to pass through: they are allowed sometimes to get choked up with dirt. Sheets of lead or copper have been used for that purpose, but they are very expensive. Asphalted felt is quite as good; no damp can pass through it. Care must, however, be taken in using it if only one wall, say a party wall, has to be built. To lay two or three courses of slates, bedded in cement, is a good method, providing the slates "break joint," and are well bedded in the cement. Workmen require watching while this is being done, because if any opening be left for damp to rise, it will undoubtedly do so. A better method is to build brickwork a few courses in height with Portland cement instead of common mortar, and upon the upper course to lay a bed of cement of about one inch in thickness; or a layer of asphalte (providing the walls are all carried up to the same level before the asphalte is applied hot). As moisture does not penetrate these substances, they are excellent materials for keeping out wet; and it can easily be seen if the mineral asphalte has been properly applied. To keep out the damp from basement floors, lay down cement concrete 6 inches thick, and on the top, asphalte 1 inch thick, and then lay the sleepers and joists above; or bed the floor boards on the asphalte.

The walls and principal timbers of a building should always be left for some time to dry after it is covered in. This drying is of the greatest benefit to the work, particularly the drying of the walls; and it also allows time for the timbers to get settled to their proper bearings, which prevents after-settlements and cracks in the finished plastering. It is sometimes said that it is useful because it allows the timber more time to season; but when the carpenter considers that it is from the ends of the timber that much of its moisture evaporates, he will see the impropriety of leaving it to season after it is framed, and also the cause of framed timbers of unseasoned wood failing at the joints sooner than in any other place. No parts of timber require the perfect extraction of the sap so much as those that are to be joined.

When the plastering is finished, a considerable time should be allowed for the work to get dry again before the skirtings, the floors, and other joiners' work be fixed. Drying will be much accelerated by a free admission of air, particularly in favourable weather. When a building is thoroughly dried at first, openings for the admission of fresh air are not necessary *when the precautions against any new accessions of moisture have been effectual.* Indeed, such openings only afford harbour for vermin: unfortunately, however, buildings are so rarely dried when first built, that air-bricks, &c., in the floors are very necessary, and if the timbers were so dried as to be free from water (which could be done by an artificial process), the wood would only be fit for joinery purposes. Few of our readers would imagine that water forms $\frac{1}{5}$th part of wood. Here is a table (compiled from 'Box on Heat,' and Péclet's great work 'Traité de la Chaleur'):—

WOOD.

Elements.	Ordinary state.
Carbon	·408
Hydrogen	·042
Oxygen	·334
Water	·200
Ashes	·016
	1·000

Many houses at our seaport towns are erected with mortar, having sea-sand in its composition, and then dry rot makes its appearance. If no other sand can be obtained, the best way is to have it washed at least *three times* (the contractor being under strict supervision, and subject to heavy penalties for evasion). After each washing it should be left exposed to the action of the sun, wind, and rain: the sand should also be frequently turned over, so that the whole of it may in turn be exposed; even then it tastes saltish, after the third operation. A friend of ours has a house at Worthing, which was erected a few years since with sea-sand mortar, and on a wet day there is always a dampness hanging about the house—every third year the staircase walls have to be repapered: it "bags" from the walls.

In floors next the ground we cannot easily prevent the access of damp, but this should be guarded against as far as possible. All mould should be carefully removed, and, if the situation admits of it, a considerable thickness of dry materials, such as brickbats, dry ashes, broken glass, clean pebbles, concrete, or the refuse of vitriol-works; but no lime (unless unslaked) should be laid under the floor, and over these a coat of smiths' ashes, or of pyrites, where they can be procured. The timber for the joists should be well seasoned; and it is advisable to cut off all connection between wooden ground floors and the rest of the woodwork of the building. A flue carried up in the wall next the kitchen chimney, commencing under the floor, and terminating at the top of the wall, and covered to prevent the rain entering, would take away the damp under a kitchen floor. In Hamburg it is a common practice to apply mineral asphalte to the basement floors of houses to prevent capillary attraction; and in the towns of the north of France, gas-tar has become of very general use to protect the basement of the houses from the effects of the external damp.

Many houses in the suburbs (particularly STUCCONIA) of London are erected by speculating builders. As soon as the carcase of a house is finished (perhaps before) the builder is unable to proceed, for want of money, and the carcase is allowed to stand unfinished for months. Showers of rain saturate the previously unseasoned timbers, and pools of water collect on the basement ground, into which they gradually, but surely, soak. Eventually the houses are finished (probably by half a dozen different tradesmen, employed by a mortgagee); bits of wood, rotten sawdust, shavings, &c., being left under the basement floor. The house when finished, having pretty (!) paper on the walls, plate-glass in the window-sashes, and a bran new brick and stucco portico to the front door, is quickly let. Dry rot soon appears, accompanied with its companions, the many-coloured fungi; and when their presence should be known from their smell, the anxious wife probably exclaims to her husband, "My dear! there is a very strange smell which appears to come from the children's playroom: had you not better send for Mr. Wideawake, the builder, for I am sure there is *something the matter with the drains.*" Defective ventilation, dry rot, *green* water thrown down sinks, &c., do not cause smells, it's *the drains,* of course!

There is another cause which affects all wood most materially, which is the application of paint, tar, or pitch before the wood has been thoroughly dried. The nature of these bodies prevents all evaporation; and the result of this is that the centre of the wood is transformed into touchwood. On the other hand, the doors, pews, and carved work of many old churches have never been painted, and yet they are often found to be perfectly sound, after having existed more than a century. In Chester, Exeter, and other old cities, where much timber was formerly used, even for the external parts of buildings, it appears to be sound and perfect, though black with age, and has never been painted.

Mr. Semple, in his treatise on 'Building in Water,' mentions an instance of some field-gates made of home fir, part of which, being near the mansion, were painted; while the rest, being in distant parts of the grounds, were not painted. Those which were painted soon became quite rotten, but the others, which were not painted, continued sound.

Another cause of dry rot, which is sometimes found in suburban and country houses, is the presence of large trees near the house. We are acquainted with the following remarkable instance:—At the northern end of Kilburn, London, stands Stanmore Cottage, erected a great many years ago: about fifty feet in front of it is an old elm-tree. The owner, a few years since, noticed cracks round the portico of the house; these cracks gradually increased in size, and other cracks appeared in the window arches, and in different parts of the external and internal walls. The owner became alarmed, and sent for an experienced builder, who advised underpinning the walls. Workmen immediately commenced to remove the ground from the foundations, and it was then found that the foundations, as well as the joists, were honeycombed by the roots of the elm-tree, which were growing alongside the joists, the whole being surrounded by large masses of white and yellow dry-rot fungus.

The insufficient use of tarpaulins is another frequent cause of dry rot. A London architect had (a few years since) to superintend the erection of a church in the south-west part of London; an experienced builder was employed. The materials were of the best description and quality. When the walls were sufficiently advanced to receive the roof, rain set in; as the clown in one of Shakespeare's plays observed, "the rain, it raineth every day;" it was so, we are told, in this case for some days. The roof when finished was ceiled below with a plaster ceiling; and above (not with "dry oakum without pitch" but) with slates. A few months afterwards some of the slates had to be reinstated, in consequence of a heavy storm, and it was then discovered that nearly all the timbers of the roof were affected by dry rot. This was an air-tight roof.

In situations favourable to rot, painting prevents every degree of exhalation, depriving at the same time the wood of the influence of the air, and the moisture runs through it, and insidiously destroys the wood. Most surveyors know that moist oak cills to window frames will soon rot, and the painting is frequently renewed; a few taps with a two-feet brass rule joint on the top and front of cill will soon prove their condition. Wood should be a year or more before it is painted; or, better still, never painted at all. Artificers can tell by the sound of any substance whether it be healthy or decayed as accurately as a musician can distinguish his notes: thus, a bricklayer strikes the wall with his crow, and a carpenter a piece of timber with his hammer. The Austrians used formerly to try the goodness of the timber for shipbuilding by the following method: One person applies his ear to the centre of one end of the timber, while another, with a key, hits the other end with a gentle stroke. If the wood be sound and good, the stroke will be distinctly heard at the other end, though the timber should be fifty feet or more in length. Timber affected with rot yields a particular sound when struck, but if it were painted, and the distemper had made much progress, with no severe stroke the outside breaks like a shell. The auger is a very useful instrument for testing wood; the wood or sawdust it brings out can be judged by its smell; which may be the fresh smell of pure wood: the vinous smell, or first degree of fermentation, which is alcoholic; or the second degree, which is putrid. The sawdust may also be tested by rubbing it between the fingers.

According to Colonel Berrien, the Michigan Central Railroad Bridge, at Niles, was painted *before seasoning*, with "Ohio fire-proof paint," forming a glazed surface. After five years it was so rotten as to require rebuilding.

Painted floor-cloths are very injurious to wooden floors, and frequently produce rottenness in the floors that are covered with them, as the painted cloth prevents the access of air, and retains whatever dampness the boards may absorb, and therefore soon causes decay. Carpets are not so injurious, but still assist in retarding free evaporation.

Captain E. M. Shaw, in 'Fire Surveys,' thus writes of the floors of a building, "They might with advantage be caulked like a ship's deck, only with dry oakum, without pitch." Let us see how far oil floor-cloth and kamptulicon will assist us in obtaining an air-tight floor.

In London houses there is generally *one* room on the basement floor which is carefully covered over with an oiled floor-cloth. In such a room the dry rot often makes its appearance. The wood absorbs the aqueous vapour which the oil-cloth will not allow to escape; and being assisted by the heat of the air in such apartments, the decay goes on rapidly. Sometimes, however, the dry rot is only confined to the top of the floor. At No. 106, Fenchurch Street, London, a wood floor was washed (a few years since) for a tenant, and oil-cloth was laid down. Circumstances necessitated his removal a few months afterwards; and it was then found that the oil-cloth had grown, so to speak, to the wood flooring, and had to be taken off with a chisel: the dry rot had been engendered merely on the surface of the floor boards, as they were sound below as well as the joists: air bricks were in the front wall.

We have seen many instances of dry rot in *passages*, where oiled floorcloth has been nailed down and not been disturbed for two or three years.

In ordinary houses, where floorcloth is laid down in the

front kitchen, no ventilation under the floors, and a fire burning every day in the stove, dry rot often appears. In the *back* kitchen, where there is no floorcloth, and only an occasional fire, it rarely appears. The air is warm and stagnant under one floor, and cold and stagnant under the other: at the temperature of 32° to 40° the progress of dry rot is very slow.

And how does *kamptulicon* behave itself? The following instances of the rapid progress of dry rot from external circumstances have recently been communicated to us; they show that, under favourable circumstances as to choice of timber and seasoning, this fungus growth can be readily produced by *casing-in* the timber with substances impervious, or nearly so, to air.

At No. 29, Mincing Lane, London, in *two* out of three rooms on the *first* floor, upon a fire-proof floor constructed on the Fox and Barrett principle (of iron joists and concrete with yellow pine sleepers, on strips of wood bedded in cement, to which were nailed the yellow pine floor-boards) kamptulicon was nailed down by the tenant's orders. In less than nine months the whole of the wood sleepers, and strips of wood, as well as the boards, were seriously injured by dry rot; whilst the third room floor, which had been covered with a carpet, was perfectly sound.

At No. 79, Gracechurch Street, London, a room on the second floor was inhabited, as soon as finished, by a tenant who had kamptulicon laid down. This floor was formed in the ordinary way, with the usual sound boarding of strips of wood, and concrete two inches thick filled in on the same, leaving a space of about two inches under the floor boards. The floor was seriously decayed by dry rot in a few months down to the level of the concrete pugging, below which it remained sound, and could be pulled up with the hand.

We will now leave oil-cloth and kamptulicon, and try what "Keene's cement" will do for an "air-tight" partition of a house.

At No. 16, Mark Lane, London, a partition was constructed of sound yellow deal quarters, covered externally with "Keene's cement, on lath, both sides." It was removed about two years after its construction, when it was found that the timber was completely *perished from dry rot;* so much so, that the timbers parted in the middle in places, and were for some time afterwards moist.

It is still unfortunately the custom to keep up the old absurd fashion of disguising woods, instead of revealing their natural beauties. Instead of wasting time in perfect imitations of scarce or dear woods, it would be much better to employ the same amount of time in fully developing the natural characteristics of many of our native woods, now destined for decorative purposes because they are cheap and common; although many of our very commonest woods are very beautifully grained, but their excellences for ornamentation are lost because our decorators have not studied the best mode of developing their beauties. Who would wish that stained deal should be painted in imitation of oak? or that the other materials of a less costly and inferior order should have been painted over instead of their natural faces being exposed to view?

There are beauties in all the materials used. The inferior serve to set off by comparison the more costly, and increase their effect. The red, yellow, and white veins of the pine timber are beautiful: the shavings are like silk ribbons, which only nature could vein after that fashion, and to imitate which would puzzle all the *tapissiers* of the Rue Mouffetard, in Paris.

Why should not light and dark woods be commonly used in combination with each other in our joinery? Wood may be stained of various shades, from light to dark. The dirt or dust does not show more on stained wood than it does on paint, and can be as easily cleaned and refreshed by periodical coats of varnish. Those parts subjected to constant wear and tear can be protected by more durable materials, such as finger-plates, &c. Oak can be stained dark, almost black, by means of bichromate of potash diluted with water. Wash the wood over with a solution of gallic acid of any required strength, and allow it to thoroughly dry. To complete the process, wash with a solution of iron in the form of "tincture of steel," or a decoction of vinegar and iron filings, and a deep and good stain will be the result. If a positive black is required, wash the wood over with gallic acid and water two or three times, allowing it to dry between every coat; the staining with the iron solution may be repeated. Raw linseed oil will stay the darker process at any stage.

Doors made up of light deal, and varied in the staining, would look as well as the ordinary graining. Good and well-seasoned materials would have to be used, and the joiners' work well fitted and constructed. Mouldings of a superior character, and in some cases gilt, might be used in the panels, &c. For doors, plain oak should be used for the stiles and rails, and pollard oak for the panels. If rose-wood or satin-wood be used, the straight-grained wood is the best adapted for stiles and rails; and for mahogany doors, the lights and shades in the panels should be stronger than in the stiles and rails.

Dark and durable woods might be used in parts most exposed to wear and tear.

Treads of stairs might be framed with oak nosings, if not at first, at least when necessary to repair the nosings.

Skirtings could be varied by using dark and hard woods for the lower part or plinth, lighter wood above, and finished with superior mouldings. It must, however, be remembered that, contrary to the rule that holds good with regard to most substances, the colours of the generality of woods become considerably *darker* by exposure to the light; allowance would therefore have to be made for this. All the woodwork *must*, previously to being fixed, be well seasoned.

The practice here recommended would be more expensive than the common method of painting, but in many cases it would be better than graining, and cheaper in the long run. Oak wainscot and Honduras mahogany doors are twice the price of deal doors; Spanish mahogany three times the price. When we consider that by using the natural woods, French polished, we save the cost of four coats of paint and graining (the customary modes), the difference in price is very small. An extra 50*l*. laid out on a 500*l*. house would give some rooms varnished and rubbed fittings, without paint. Would it not be worth the outlay? It may be said that spots of grease and stains would soon disfigure the bare wood; if so, they could easily be removed by the following process: Take a quarter of a pound of fuller's earth, and a quarter of a pound of pearlash, and boil them in a quart of soft water, and, while hot, lay the composition on the greased parts, allowing it to remain on them for ten or twelve hours; after which it may be washed off with fine sand and water. If a floor be much spotted with grease, it should be completely washed over with this mixture, and allowed to remain for twenty-four hours before it is removed.

Let us consider how we paint our doors, cupboards, &c., at the present time. For our best houses, the stiles of our doors are painted French white; and the panels, pink, or salmon colour! For cheaper houses, the doors, cupboards, window linings, &c., are generally two shades of what is called "stone colour" (as if stone was always the same colour), and badly executed into the bargain: the best rooms having the woodwork grained in imitation of oak, or satin-wood, &c. And such imitations! Mahogany and oak are now even imitated on leather and paper-hangings. Wood, well and cleanly varnished, stained, or, better still, French polished, must surely look better than these daubs. But French polish is not extensively used in England: it is confined to cabinet pieces and furniture, except in the houses of the aristocracy. Clean, colourless varnish ought to be more generally used to finish off our woodwork, instead of the painting now so common. The varnish should be clean and colourless, as the yellow colour of the ordinary varnishes greatly interferes with the tints of the light woods.

In the Imperial Palace, at Berlin, one or two of the Emperor's private rooms are entirely fitted up with deal fittings; doors, windows, shutters, and everything else of fir-wood. "Common deal," if well selected, is beautiful, cheap, and pleasing.

We have seen the offices of Herr Krauss (architect to Prince and Princess Louis of Hesse), who resides at Mayence, and they are fitted up, or rather the walls and ceilings are lined, with picked pitch pine-wood, parts being carved, and the whole French polished, and the effect is much superior to any paint, be it "stone colour," "salmon colour," or even "French white."

The reception-room, where the Emperor of Germany usually transacts business with his ministers, and receives deputations, &c., as well as the adjoining cabinets, are fitted with deal, not grained and painted, but well French polished. The wood is, of course, carefully selected, carefully wrought, and excellently French polished, which is the great secret of the business. In France, it is a very common practice to polish and wax floors.

The late Sir Anthony Carlisle had the interior woodwork of his house, in Langham Place, London, varnished throughout, and the effect of the varnished deal was very like satin-wood.

About forty years since, Mr. J. G. Crace, when engaged on the decoration of the Duke of Hamilton's house, in the Isle of Arran, found the woodwork of red pine, so free from knots, and so well executed, that instead of painting it, he had it only varnished. It was a great success, and ten years after looked nearly as well as when first done.

The late Mr. Owen Jones, whose works on colour decoration are well known, was employed a few years since by Mr. Alfred Morrisson to decorate his town and country houses. At the country house (Fonthill House), Mr. Jones built a room for the display of Chinese egg-shell pottery, the chimneypiece and fittings being entirely of ebony, inlaid with ivory, and the ceiling of wood, panelled and inlaid, the mouldings being black and gold. At the town house, in Carlton House Terrace, London, the woodwork of the panelling, dadoes, doors, architraves, window-shutters, and all the rooms on the ground and first floors is inlaid, from designs by Mr. Jones, with various woods of different kinds, the colours of which were carefully selected by him, with a view to perfect harmony of colouring.

A house has recently been erected (from the designs of Mr. J. W. McLaughlin, architect) near Cincinnati, Ohio, United States, which is a perfect model with regard to the amount of woodwork used. The walls of the *hall* are finished with walnut wainscoting; the fireplace is an open one, with a walnut mantelpiece, surmounted by three statues, Peace, Plenty, and Harmony, supporting the carved wooden cornice. The Elizabethan *staircase* has carved panels of maple. The *library* is wainscoted to the ceiling in black walnut, inlaid with ebony. The *dining room* is also wainscoted in the richest style in oak, with polished mahogany panels. The *floors* are of marquetry, of different woods and patterns. The *chamber story* is finished in oak and walnut, with mahogany in the panels. The *entire interior* finish of the house is of hard wood, *varnished and rubbed* in cabinet style. This is as it should be for a gentleman's residence.

We believe the largest house now being erected in London is from the designs of Mr. Knowles, jun., for Baron Albert Grant, at Kensington. We have not seen it, but we hope it will be finished in the Cincinnati style, as far as regards the amount of ornamental woods used.

There is a cynical French proverb, which says, " When we cannot have what we love, we must love what we have." But surely this cynical proverb cannot be applied to "stone colour" paint on wood. The Japanese, however, some years since, determined not to follow this advice, for when the English Government, at Admiral Sterling's suggestion, sent to the Tycoon a very fine steam vessel, the Japanese (who abhor paint about their ships) immediately commenced to scrub off the paint. According to Sir Rutherford Alcock, they have been steadily engaged in scrubbing it off ever since the boat has come into their possession, and by dint of labour and perseverance have nearly succeeded. All the fine imitation satin-wood and the gilt work have been reduced to a very forlorn state. The Japanese not only decline to follow advice, but they are a very difficult race of people from whom to obtain correct information. When Mr. Veitch was at Yeddo, on a visit to the Legation, in quest of botanical specimens, he saw a pine-tree from which he desired a few seeds. "Oh," said the inevitable yaconins, "those trees *have no seed!*"—" But there they are," replied the unreasonable botanist, pointing to some. "Ah, yes, true; but they *will not grow,*" was the reply.

If we must take our fashions from royalty and the aristocracy, and if we must go abroad for them, surely the above examples will suffice; but if we must have paint, then the preservative solution, now being extensively used in the restoration and renovation of St. Paul's Cathedral, under the superintendence of Mr. F. C. Penrose, the architect to the Dean and Chapter, appears to possess several good qualities. The preservative solution, which is manufactured by the Indestructible Paint Company, is said to be as follows: 1st, that it is colourless and invisible; 2nd, in no way does it alter the appearance of the surface; 3rd, it prevents the growth of vegetation; and 4th, that it resists the action of the atmosphere and changes of weather, not only preventing but also arresting decay.

It is necessary that the wood selected (if not to be painted) should be well grown, and from a fully developed tree, where all the fibres or grain are distinctly marked. The beauty of the wood, when properly treated, consists in the brilliant manner in which the rich, deep yellow streaks or layers of the hard wood are developed under the hands of the skilful polisher. These yellow veins show through the polish like clear and beautifully marked streaks of amber; and strongly reflecting the light, they produce a very pleasing effect. The yellow, variegated, hard part of the wood forms a very excellent contrast to the delicate whiteness of the softer parts of the board; and, if skilfully selected, the effect will be much admired, and certainly preferred to the best imitation of the more rare and expensive woods. In arranging doors, panels, &c., much will, of course, depend in selecting the wood, in placing the best parts in the panels, so that when polished the most pleasing effects will be produced. Much, too, depends on skilful workmanship and smooth finish, which can only be obtained by care, and using well-seasoned wood; but this is the case with all species of wood.

Should any young architect, after reading the preceding remarks, be desirous of employing natural woods in his building works, we advise him, before he attempts this kind of colour decoration, to study Mr. Owen Jones' lecture on "Colour in the Decorative Arts," delivered before the Society of Arts, 1852; and likewise M. Chevreul's 'Laws of the Simultaneous Contrast of Colours'; we also recommend him to—

Use *moderate* things elegantly, and *elegant* things moderately.

Oak, walnut, maple, elm, and some other woods become of very dark colour, but can be made to receive a fine polish, and could often be employed for panels with good effect. In some cases there is great contrast of tint in the same log after preparation, so that these might be inapplicable except in smaller pieces, or perhaps by applying the process after the work has been made; but sycamore, beech, and some other woods are generally uniform, except as regards the previous grain of the wood.

As to the matter of showing the end of the grain, according to the Gothic principle the beauty of a wood consists in showing the end of the grain; but, at the same time, the classic principle is that there is a greater beauty in the side way of the grain than in the end way.

Although varnish and polish both form a glazing, and give a lustre to the wood they cover, as well as heighten the colours of the wood, yet from their want of consistence they are liable to yield to any shrinking or swelling, rising in scales or cracking, when much knocked about. Waxing, on the contrary, resists percussion, but it does not possess in the same degree as varnish the property of giving lustre to the bodies on which it is applied; any accidents, however, to its polish are easily repaired by rubbing.

The woodwork of the Swiss Cottage, at the late Colosseum, London, in the Regent's Park, was only varnished.

In using stain on any description of wood, the stain should always be allowed to get quite dry before sizing, as that gives it a fair chance of striking into the wood. Glue size is the best for stained work, made so thin that there is no fear of putting it on in patches. After the size is quite dry also, varnish; and if the first coat does not stand out quite sufficiently to please the eye, give it a second coat. Some persons use stain and varnish together, doing away with size; but this is a very poor method, for should the wood get scratched or damaged in any way, the varnish and stain come off together, leaving a white place, if it be white wood that is stained. A painter who has been in the trade forty years, recently remarked to us, "You must size, or else the varnish won't come out; it won't show that it *is* varnish; the wood soaks it up; while there is any suction going on the varnish 'll go in. The sizing stops all suction."

A great many experiments and attempts have been made at different times to colour wood. John of Verona first conceived the idea. The celebrated B. Pallissy investigated the cause of the veins, &c., in wood, and tried mordant solutions applied to the surface, wetting the surface with certain acids, immersing the wood in water to bring out the veinage, &c.

Ebony has often been imitated by penetrating sycamore, plane, and lime woods to a certain depth with pyrolignite of iron, gall-nuts, &c.

Werner, in 1812, obtained great success at Dijon in colouring the woods by filtration. Marloye, in 1833, constructed a machine to colour wood by placing it erect in a cylinder, sucking out the air at one end, and forcing up the colouring solution through the other. He gave the credit of this to Bréant. Marloye has manufactured many mathematical instruments of wood coloured in this way, *which does not warp.*

If we could afford the space, we would willingly give a *résumé* of the attempts of well known experimentalists to colour wood. We can only give the year and name in each case:

1709.	Magnol.	1735.	Buffon.
1733.	La Baisse.	1754.	Bonnet.
1735.	Hales.	1758.	Du Hamel.
	1804.	Saussure.	

During the recent war between France and Germany, the latter country advanced matters, their supplies of coloured woods from France being gone.

As we have made so many remarks against painting

wood, it is only right that we should give some description of it, which we will now do.

House painting, according to Mr. W. Papworth, in his lecture on "Fir, Deal, and House Painting," 1857, did not come into *general* use until about the period of William and Mary, and Anne, up to which time either colouring by distemper or by whitewash had been in vogue for plaster work, leaving inside woodwork more or less untouched.

We think, without wishing to think *too loud*, that house painting was invented by *a bad builder*, in the seventeenth century, because

Putty and paint cover a multitude of sins.

The process of graining and marbling may be traced back as far at least as the time of James III. of Scotland (1567-1603), during whose reign a room of Hopetown Tower was painted in imitation of marble. Before that period, imitations were done in "stone" colour, "marble" colour, wainscot colour, &c. In 1676, marbling was executed as well as imitations of olive and walnut woods; and in 1688 tortoise-shell was copied on battens and mouldings. Maghogany was imitated in 1815, and maple wood in 1817. But why imitate mahogany, when the grain of the wood differs so much in texture, and in the appearance of the different and beautiful shades, technically termed *roe, broken roe,* **bold roe,** *mottle, faint mottle,* and *dapple.*

The following description will give the reader some idea of ordinary painting. The woodwork having been prepared for fixing, has first to undergo the process of "knotting," in order to prevent the turpentine in the knots of fir-wood from passing through the several coats of paint. One method for best work is to cut out the knot whilst the work is at the bench to a slight depth, and to fill up the hole with a stiff putty made of white lead, japan, and turpentine. There are many ways of killing the knots: the best and surest is to cover them with gold or silver leaf. Sometimes a lump of fresh slaked lime is laid on for about twenty-four hours, then scraped off, a coating of "size knotting" applied, and if not sufficiently killed, they are coated with red and white lead in linseed oil, and rubbed down when dry. The general method is to cover the parts with size knotting, which is a preparation of red lead, white lead, and whitening, made into a thin paste with size. The most common mode is to paint them with red ochre, which is worth nothing. The next process is that of priming, which consists in giving a coat of white and red lead, and a little dryers in linseed oil. This is the first coat, and upon which the look of the paint on completion depends. This first, or priming coat, is put on before "stopping" the work, should that process be required. It consists in filling up with putty any cracks or other imperfections on the surface of the wood. If the putty used in the process of stopping be introduced before the first coat of colour is laid on, it becomes loose when dry. After this first coat, pumicing is resorted to for removing all irregularities from the surface. It is worth recollecting that *old* white lead is much superior to new for all painting operations.

A smooth surface being thus obtained, the second coat is given, consisting of white lead and oil: about one-fourth part of turpentine is sometimes added for quick work. If four coats are to be laid on, this second one has sometimes a proportion of red lead, amounting to a flesh colour; but if only three, it is generally made to assume the tint of the finishing coat. It should have a good body, and be laid even. This coat, when thoroughly dry and hard, is, in best work, rubbed down with fine sand paper, and then the third coat, or "ground colour," applied of a somewhat darker tint than wanted when finished, having sufficient oil for easy working, but not too fluid: thus two-thirds oil, and one-third turpentine. The "flatting" coat follows, the object of which is to prevent the gloss or glaze of the oil, and to obtain a flat, dead appearance. White lead is mixed with turpentine, to which a little copal is sometimes added, and when the tint is put in it is always made lighter than the ground colour, or it would, when finished, appear in a series of shades and stripes. Flatting must be executed quickly, and the brush is generally, if not always, carried up the work, and not across it.

To clean paint, raw alkalies should not be used, as they will infallibly take off the flatting coat. The best mode of cleaning is by means of good soap, not too strong, laid on with a large brush, so as to make a lather: this should be washed off clean with a sponge, and wiped dry with a leather.

We must draw to a conclusion.

One cause of the decay of modern buildings, and frequent cases of dry rot, is owing to the employment of bad builders. We advise the non-professional reader to employ an architect or surveyor when he desires to speculate in bricks and mortar: it is the cheapest course. If he doubts the truth of what we have written, we can assure him he will be a mere child in the hands of a bad or scamping builder; that is to say, he will obtain a badly-erected house,—a cheap contract, and a long bill of extras.

There are *seven* classes of bad builders—1st, the *bad builder* who does not know his business; 2nd, the *bad builder* who has no money to carry it on with; 3rd, the *partial* scamp; 4th, the *regular* scamp; 5th, the *thorough* scamp; 6th, the *"jerry"* builder; and 7th, the *vagabond*. There is an instance of the latter class given by Mr. Menzies in his fine work on 'Windsor Park,' 1864. We could give examples of all these classes, and draw the line between each class, impossible as it may seem: they are always looking out for customers, *without architects*.

We could assist the non-professional reader by quoting the advice given by several architects (viz. Sir C. Wren, C. Barry, R. Smirke, W. Chambers, and W. Tite) relative to buildings, but there is a Danish proverb which, translated into English, runs as follows: "He who builds according to every man's advice will have a crooked house."

CHAPTER VII.

ON THE PRESERVATION OF WOODEN BRIDGES, JETTIES, PILES, HARBOUR WORKS, ETC., FROM THE RAVAGES OF THE TEREDO NAVALIS AND OTHER SEA-WORMS.

> "Perforated sore
> And drilled in holes, the solid oak is found
> By worms voracious, eaten through and through."
>
> SIR JOHN BARROW.

As the destruction of timber by fungi has been called the *vegetable* rot, it may not be inappropriate to term the destruction of wood by various worms and insects, the *animal* rot.

We have *four natural enemies* to deal with: 1st, the *dry rot*, that attacks our houses, &c.; 2nd, the *worms*, or boring animals, which destroy our ships and harbours; 3rd, the *rust*, that eats our iron; and 4th, the *moisture and gases*, that destroy our stone.

There are three classes of destructive insects which prey upon timber trees, founded upon the manner in which they carry on their operations—viz. those which feed upon the leaves and tender shoots; those which feed upon the bark and the albumen; and those which feed upon the heart-wood.

It is to be observed that some of the insects which feed upon the heart of wood do not cease their ravages upon the removal of the tree; but that, on the contrary, the *Cossus syrex*, of our indigenous fauna, and the larvæ of the *Callidium bajutum*, which are often found in imported timber, continue to devour the wood long after it has been inserted in buildings. There seem to be very few means of defence against this class of destructive agents; and very few trustworthy indications of their existence, or of the extent of the ravages they have committed, are to be discovered externally; and it thus frequently happens that a sound, hearty-looking stick of timber may be so seriously bored by these insects as to be of comparatively little value for building purposes of any description. The soft and tender woods, and such as are of a saccharine nature in their juices, are the most liable to be assailed by worms; those which are bitter are generally, if not invariably, exempt; it is obvious, therefore, that those palatable juices, which are so conducive to their production and propagation, should be got rid of by thorough seasoning, and, if further precaution be necessary, that the infusion of some bitter decoction into the pores of the wood will be an effectual preventive; and for which those woods that are of a regular grain afford sufficient facilities. Ash, if felled when abounding in sap, is very subject to worms; beech, under similar circumstances, is also liable to their attacks; likewise alder and birch; in these woods water seasoning is sometimes found to be a good preventive; the sapwood of oak is also thus improved; the silver fir is subject to them; the sycamore is rather so; alder is said when dry to be very susceptible of engendering them; the cedar, walnut, plane, cypress, and mahogany are examples of woods which discourage their advances. It has been stated that Robert Stevenson (not the son of the "Father of Railways"), of Edinburgh, at Bell Rock Lighthouse (of which he was engineer), between 1814 and 1843, found that greenheart wood, beef wood, and bullet tree were not perforated by the *Teredo navalis*, and teak but slightly so. Later experiments show that the "jarrah" of the East, also, is not attacked. Lignum vitæ is said to be exempt. The cost of these woods prevents their general use.

In 1810, Stevenson first noticed the *teredo* in piles, and specimens of the creatures in wood were sent to Dr. Leach, of the British Museum, in 1811, who examined them, and noticed their peculiarities. Stevenson, settled on Bell Rock during many years (like a new Robinson Crusoe), was enabled to watch the injuries done to the piles by the *teredo*. With piles which had been subjected to Kyan's process before immersion, the wood was attacked at the end of the twenty-eighth month, and was entirely destroyed in the seventh month of the fifth year. With Payne's, it lasted a year longer.

We can give the names of those who have given much time and attention to this subject. At the bottom of this page a list of works of reference * will be found useful. Messrs. Stevenson (engineer of Bell Rock Lighthouse), Harting (Member of the Academy of Sciences of the

---

* See 'Proceedings of the Royal Society of Edinburgh,' v. 7, page 433; 'Tredgold's Carpentry,' by J. T. Hurst, 1871; 'Histoire de l'Acad.,' 1765, page 15; 'Ann. des Ponts et Chaussées,' v. 15, page 307; 'Mem. sur la Conservation des Bois à la Mer,' 1868, by Forestier; 'Bois de Marine,' by Quatrefages, 1848.

Pays-Bas), de Quatrafages, Deshayes, Caillaut, Hancock, Dagneau, de Gemini, Kater, Crepin (Engineer-in-Chief, of Belgium), and A. Forestier (Engineer-in-Chief of the Bridges, &c., of France).

The termite, or white ant, is the most destructive insect to timber on land, whilst the teredo reigns supreme of sea worms in the sea. The former we shall treat of in our next chapter, the latter we propose considering at some length in this.

The marine worm, of which there are accounts in all parts of the world, has been known, by its effects, for hundreds of years; indeed, Ovid spoke of it nineteen hundred years ago, and it is even mentioned by Homer. Fossil *terredines* of great antiquity have been found near Southend; also pieces of petrified wood from the greensand, near Lyme and Sidmouth, bored by ancient species of *teredo*; also from Bath, and from Doulting, near Shepton Mallet, specimens of oolite, with petrified corallines in it, pierced by boring shells.

It is said that this worm is a native of India, and that it was introduced to Holland some 200 years ago, from whence it has spread through the ports of northern Europe.

The *Teredo navalis** is very destructive to harbour works and piling. The Southampton water is particularly infested with it; in fact, the *teredo* is found in every port to which coals are carried south of the Tees; in the Thames, as high up as Gravesend; and northward as far

---

* There are eight kinds of *teredines*, of which three are to be found in European waters, viz. the *Teredo fatalis*, *Teredo navalis*, *Teredo bipennata*.

as Whitby. It is also found at Ryde, Brighton, and Dover. Traces of the ravages of the *Teredo navalis*, and of the *Limnoria terebrans*, have at various periods been found from the north of Scotland and Ireland, on almost every coast, to the Cape of Good Hope and Van Dieman's Land, in the eastern hemisphere; and, in the western hemisphere, from the river St. Lawrence to Staten Island, near Terra del Fuego, almost in the Polar Sea; so that although this maritime scourge is rifest in warm climates, yet cold latitudes are not exempt from it.

At the Crystal Palace, Sydenham, may be seen the destructive *Teredo navalis* in a bottle, and there may also be seen mahogany perforated by it, and fir piles from Lowestoft Harbour, which were rendered useless by the ravages of the worm and the *limnoria* three years after they were driven, showing the necessity of defending timber intended for marine construction. A specimen of American oak from the dock gates of Lowestoft Harbour, which had been four years under water, and a part of a fir-pile from the dockyard creek at Sevastopol, also show the destructive powers of the *teredo*. At the South Kensington and British Museums, London, specimens of this worm may also be seen, as well as pieces of timber perforated by it.

The bottoms of ships, and timbers exposed to the action of the sea, are often destroyed by the *teredo*.

The gunboats constructed during the Crimean war suffered far more from dry rot and the *teredo* than the shot and shell of the Russians. One cannot even guess at the mischief perpetrated every year all along our shores, in docks and harbours, by the boring animals that penetrate all woods not specially protected. We cannot count the number of the ships that have foundered at sea, owing to those few inches of timber, on which all depended, being pierced or destroyed by the worm or fungus.

In the short space of twelve years these destructive worms were known to make such havoc in the fir piles of a bridge at Teignmouth, that the whole bridge fell suddenly, and had to be totally reconstructed.

The wooden piers of Bridlington were nearly wholly destroyed by worms; and the pile fenders on the stone piers at Scarborough were generally cut through in a few years.

At Dunkirk, wooden jetties are so speedily eaten away that they require renewal every twelve or fifteen years. At Havre, a stockade was entirely destroyed in six months. At Lorient, wood only lasts about three years in the sea-water; and at Aix, the hull of a stranded vessel was found to have lost half its weight in six months, from the ravages of these animals.

The reason why Balaclava, in Russia, is not a place of considerable mercantile importance is owing in a great measure to the destructive ravages of the worms with which its waters are infested, and by which the hulls of ships remaining there for any length of time become perforated.

The piles of the jetties in Colombo Harbour, Ceylon, which are mostly of satinwood, and about 14 inches in diameter, are so pierced by these worms in the course of twelve months as to require renewal.

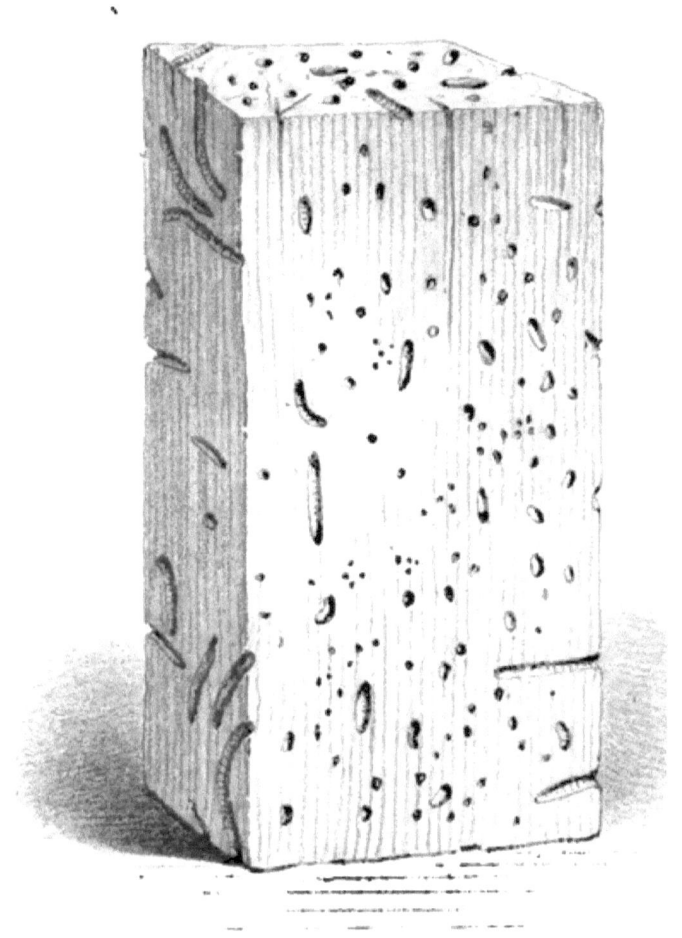

Portion of pile, from Balaclava harbour, Russia, 'riddled' by the Teredo Navalis.

The cofferdam at **Sheerness** was destroyed by the *teredo*. After a time, it was no uncommon occurrence to see several piles, apparently sound, floated away at each tide; indeed, they were so thoroughly perforated by the *teredo* that in still weather, by putting the ear to the side of the pile, the worms could be heard at their boring labours.

The almost total destruction of the pier-head of the old Southend Pier in a few years, is another instance of the serious damages these worms cause. The old pier-head was erected in the year 1833, and in three years the majority of the wooden piles had been almost destroyed, and at the end of ten years, in addition to the piles being all eaten through by the worms, the whole structure had sunk 9 inches at the western end, so that in a short time it would have fallen. The materials with which the work was constructed were of good quality, the fir being Memel, and the oak of English growth; it was all perfectly sound in those places were the *teredo* had not attacked it, and indeed portions of it were again used in the construction of the extension of the pier. The whole of the timber work was well coated with pitch and tar previously to being fixed, but notwithstanding these precautions, and an apparent determination to protect the pier-head by copper sheathing, brushing, cleaning, and constant watchfulness, the *teredo* made its appearance, and committed such ravages that the entire destruction of the pier-head soon appeared inevitable. The *Teredo navalis* first showed itself six months after the completion of the work, and was reported within twelve months to have seriously injured the piles above the copper, whilst at about low-water mark, of neap tides,

P

nearly all the piles exhibited appearances of destruction, the *limnoria*, as well as the *teredo*, having seriously attacked them; and in less than four years from the completion of the pier-head, they had progressed in their work to such an extent that some of the piles were entirely eaten through, both above and below the copper sheathing; in consequence of this the stability of the structure was materially injured, and, on examination, it was discovered that the ground had been considerably washed away by the action of the sea, and that the piles below the copper were exposed to the action of the *teredo*.

The first appearances of the *Teredo navalis* are somewhat singular, inasmuch as the wood which has been perforated by it presents to the casual observer no symptom of destruction on the surface, nor are the animals themselves visible, until the outer part of the wood has been broken away, when their shelly habitations come in sight, and show the perfect honeycomb they have formed; on a closer examination of the wood, however, a number of minute perforations are discovered on the surface, generally covered with a slimy matter; and on opening the wood at one of these, and tracing it, the tail of the animal is immediately found, and after various windings and turnings, the head is discovered, which, in some cases, is as much as 3 feet from the point of entrance; sometimes it will happen, especially if the wood has been much eaten, that their shelly tubes are partly visible on the surface, but this is rare; they enter at the surface, and bore in every direction, both with and against the grain of the wood, growing in size as they proceed.

The Rev. W. Wood writes, in 1863: "I have now before me a portion of the pier at Yarmouth, which is so honeycombed by this terrible creature that it can be crushed between the hands as if it were paper, and in many places the wood is not thicker than ordinary foolscap. This piece was broken off by a steamer which accidentally ran against it; and so completely is it tunnelled, that although it measures 7 inches in length and about 11 in circumference, its weight is under 4 ounces, a considerable portion of even that weight being due to the shelly tubes of the destroyers."

The eggs of the *teredo* affix themselves to the wood they are washed against, are then hatched, and the worm commences boring; each individual serves by itself for the propagation of the species; and they rarely injure each other's habitations. Any timber, constantly under water, but not exposed to the action of the air at the fall of the tide, is extremely likely to be destroyed by them. They appear to enter the wood obliquely, to take the grain of the fibre, and more generally to bore with it downwards, where the perforations are left dry at low water.

It has been stated by some authorities that the *teredo* is only a destructive creature, and seeks the wood as a shelter, from instinctive dread of some larger animals, but there is no doubt this insect feeds upon wood. Mr. John Paton, C.E. (to whom we are indebted for much information on these worms), in conjuction with Mr. Newport, the eminent physiologist and anatomist, on carefully dissecting this animal for the purpose of ascertaining its general character, and more particularly the nature of its food, found digested portions of wood in its body, so that there is no doubt that the *teredo* does feed upon the particles of the wood, and to this its rapid and extraordinary growth must be mainly attributed.

The *Teredo navalis*, or, as it is sometimes called, the Ship Worm, is one of the *Acephalous mollusca*, order Conchifera, and of the family of the *Pholadariæ*. It is of an elongated vermiform shape, the large anterior part of which constitutes the boring apparatus, and contains the organs of digestion, and the posterior, gradually diminishing in size, those of respiration. The body is covered with a transparent skin, through which the motion of the intestines and other remarkable peculiarities are plainly visible. The posterior or tail portion is armed at its extremity, with two shells, and has projecting from it a pair of tubular organs, through which the water enters, for the purpose of respiration; this portion is always in the direction of the surface, and apparently in immediate contact with the water, but does not bore. The anterior portion of the animal is that by which it penetrates the wood, being well armed for the purpose by having, on each side, a pair of strong valves, formed of two pieces, perfectly distinct from one another; the larger piece protects the sides and surface of the extremities, and has a shelly structure projecting from the interior, to which the muscles are attached; the smaller piece is more convex, and covers that part which should be regarded as the anterior surface of boring. This portion of the shell is deeply carniated, and seems to constitute the boring apparatus. The shells form an envelop around the external tegument of the animal, which even

The Lycoris, which destroys the Teredo Navalis.

Portion of Timber pile destroyed by Sea Worms.

surrounds the foot, or part by which it adheres to the wood. The neck is provided with powerful muscles. The manner in which it appears to perforate the wood is by a rotary motion of the foot, carrying round the shells, and thus making those parts act as an auger, which is kept, or retained in connection with the wood, by the strong adherence of the foot. The particles of wood removed by this continued action of the foot, and the valves, are engorged by the animal, for between the junction of the two large shells there is a longitudinal fissure in the foot, which appears to be formed by a fold of this portion of the two sides, thus forming a canal to the oral orifice, and along which the particles of wood bored out, are conveyed to the mouth. The mouth, or entrance to the digestive organs, is of a funnel shape, and consists of a soft, or membraneous surface, capable of being enlarged, and leading into an œsophagus, which passes backwards towards the dorsal surface of the animal. At or near the termination of the œsophagus, there is a glandular organ, the use of which is possibly to secrete a fluid for assisting in the digestion of the wood, and not, as has been supposed, to act as a solvent; for if such were the case, it would most probably be situated at its commencement instead of at its termination. At a short distance behind this organ are two other large glandular bodies, the use of which may also be to secrete fluid for the purpose of digestion. The œsophagus terminates in a large dilatation, into which these organs pour their contents; at its posterior end the canal is dilated into a very large elongated sac, which extends backwards to about one-fourth of the length of the whole animal, and is filled with food, while from its anterior, or upper surface, it has an oval, muscular formation, from which the alimentary canal is continued forwards, and, after making a few turns, passes backwards, in an almost direct line, on the upper surface of the large sac, again passing backwards and forwards, until it finally arrives at its termination, which it passes round, and then proceeds, in a direct line, to the anal outlet. In the lower portion of the œsophagus, and also in the sac, distinct portions of woody fibre of an extremely minute character were found by the aid of the microscope of a power of three hundred, and this was the character of the whole of the contents of the alimentary canal.

The *teredo* lines the passage in the wood with a hard shell; this shell is formed around, but does not adhere to the body; it is secreted by the external covering, which, in its first formation, is extremely fragile, but becomes hardened by contact with the water, and adheres to the wood, from which it may, however, be easily detached. The interior of this shell is not filled by the body of the *teredo*, but a large space around it is occupied with water, admitted through the small orifice in the surface of the wood through which the animal first entered; the water being drawn through the respiratory tubes, into the bronchial cavity of the body, is expired again through the same orifice, and this, in conjunction with the valve-like shells attached at this part, induces a current round the animal which removes the excreted fœtal matter. The shells are very smooth on the inner surface, but are somewhat rougher on the exterior; they are much harder and firmer in the cells of the older animals than in the young ones, and are composed of several annular parts, differing greatly in their length.

It is no less curious than wonderful to observe the mysterious instinct which apparently regulates the mechanical skill of the *teredo*, its own body supplying it with an implement of such admirable consistency and adaptation as to enable it to excavate a habitation for itself, so accurately formed that to a casual observer it would appear a mystery how so perfect a circle could be produced. It is only on examination that the raised and hollow parts of the wood become visible, and explain, in some degree, the auger-shaped contrivance that has been used for the purpose of perforating.

It has already been stated, that the wood is perforated by a rotary motion of the foot, the adhering part of which acts as a fulcrum, carrying round the shells, and thus giving immense power to the animal in its operations.

It is said that when Brunel was considering how to construct the Thames Tunnel, he was one day "passing through the dockyard (at Chatham, where he was employed by Government), when his attention was attracted to an old piece of ship-timber which had been perforated by that well-known destoyer of timber—the *Teredo navalis*. He examined the perforations, and subsequently the animal. He found it armed with a pair of strong shelly valves, which enveloped its anterior integuments; and that, with its foot as a fulcrum, a rotatory motion was given by powerful muscles to the valves, which, acting on the wood like an auger, penetrated gradually but surely; and that, as the particles were removed, they were passed through a longitudinal fissure in the foot, which formed a canal to the mouth, and so were engorged. To imitate the action of this animal became Brunel's study. 'From these ideas,' said he, 'by slow and certain methods; which, when compared with the progress of works of art, will be found to be much more expeditious in the end.'" *

Professor Owen suggests that the power of the *teredo* to bore into wood depends on muscular friction, the muscular substance being perpetually renewed while the wood wastes away, of course, without renewal. Professor Forbes, Dr. Carpenter, and Dr. Lyon Playfair were appointed about twenty-five years ago by the British Association to examine into the natural history and habits of these boring animals, but they did not arrive at any definite conclusion as to whether the boring action of the *teredo* was mechanical or chemical. Dr. Deshayes, on his return from Algiers, after making accurate drawings and careful investigations, came to the conclusion that the borings were effected by an acid secretion. Mr. Thomson, of Belfast, examined the operations of the *teredo* on the pier at Port Patrick, and arrived at the same conclusion. The general opinion, however, is that the boring action is a mechanical one.

Although the *teredo* appears to penetrate all kinds of timber, that which it seems to destroy with the greatest ease is fir, in which it works much more speedily and

---

* See 'Memoirs of Sir M. I. Brunel;' also, for particulars of the construction of the shield designed by him for forming the Tunnel, Weale's 'London Exhibited,' and 'A Memoir of the Thames Tunnel,' in Weale's Quarterly Papers on Engineering.

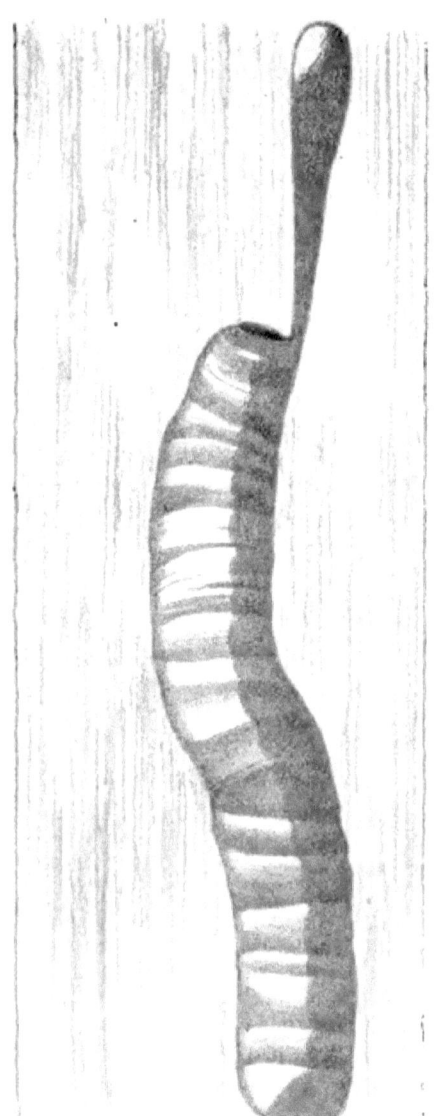

Cell formed by the Teredo Navalis showing method of boring.

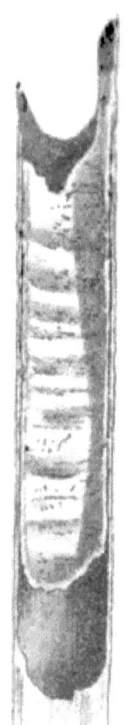

Shell left by the Teredo Navalis.

successfully than in any other, and perhaps grows to the greatest size. In a fir pile, taken from the old pier-head at Southend, a worm was found 2 feet long and ¾ inch in diameter, and indeed they have been heard of 3 feet in length and 1 inch in diameter. The soft, porous nature of the wood is no doubt the cause of their rapid growth, for in oak timber they do not progress so fast, or grow to so great a length, though in Sir Hans Sloane's 'History of Jamaica' (1725) there are accounts of these animals destroying keels of ships made of oak, and even of cedar, although the latter is renowned, by its smell and resin, for resisting all kinds of worms.

There is another kind of worm which is very destructive to timber, which Smeaton observed in Bridlington piers. This is the TIMBER-BORING SHRIMP, or GRIBBLE, the *Limnoria terebrans* (or *Limnoria perforata*, Leach), a mollusc of the family *Asselotes*, Leach. The *Limnoria terebrans* is very abundant around the British shores. Its ravages were first particularly observed in the year 1810, by the late Mr. Robert Stevenson, engineer of the Bell Rock Lighthouse. While engaged in the erection of that structure he found the timber of the temporary erections to be soon destroyed by the attacks of the *limnoria*. So little was known of the *limnoria* at the time that Dr. Leach, a well-known naturalist, who received some specimens from Mr. Stevenson, in 1811, declared it to be a new and highly interesting species. In 1834, the late Dr. John Coldstream wrote a very full and interesting description of the creature. The *limnoria* resembles a woodlouse, and is so small as hardly to be perceptible in the timber it attacks, being almost of the same colour. Small as is this crustacean, hardly larger indeed than a grain of rice, it is a sad pest wherever submarine timber is employed, for it works with great energy, and its vast numbers quite compensate for the small size of each individual; for as many as twenty thousand will appear on the surface of a piece of a pile only 12 inches square. It proceeds in a very methodical manner, and makes its way obliquely inward, unless it happens to meet a knot, when it passes round the obstacle and resumes its former direction. The surface of the timber being first attacked, it proceeds progressively into the wood to the depth of about $1\frac{1}{2}$ inch: the tunnels being cylindrical, perfectly smooth winding holes, about $\frac{1}{15}$th inch in diameter: it is necessary that the holes should be filled with salt water. The outward crust formed by these attacks then becomes macerated and rotten, and is gradually washed away by the beating of the sea. The *limnoria* does not work by means of any tool or instrument like the *teredo*, but is supposed to possess some species of dissolvent liquor, furnished by the juices of the animal itself. Dr. Coldstream was of opinion that the animal effects its work by the use of its mandibles. From ligneous matter having been found in its viscera, some have concluded that it feeds on the wood, but since other molluscs of the same genus, *Pholas*, bore and destroy stonework, the perforation may serve only for the animal's dwelling. The *limnoria* seems to prefer tender woods, but the hardest do not escape: teak and green-heart are about the only woods it does not attack. The rate at which the *limnoria* bores into wood in pure salt water is said to be about one inch in a year; but instances have occurred in which the destruction has been much more rapid. At Lowestoft Harbour, square 14 inch piles were in three years eaten down to 4 inches square. At Greenock, a pile 12 inches square was eaten through in seven years. It is stated that a 3-inch oak plank, 12 feet long, would be entirely destroyed in about eight years. Joists of timber have been found at Southend Pier, 2 feet and 3 feet below high-water mark, where they had made rapid destruction. The *limnoria* almost always works just under neap tides; it cannot live in fresh water, and whilst it is destroying the surface of a pile, the *teredo* is attacking the interior: sometimes the former is found attacking the same timber as the Chelura. As with most of these creatures, the male *limnoria* is smaller than the female, being about one-third her size. The female may be distinguished by the pouch in which the eggs and afterwards the young are carried. About six or seven young are generally found in the pouch.

The WOOD-BORING SHRIMP (*Chelura terebrans*) is a crustacean that nearly rivals the *teredo* itself in its destructive powers. It makes burrows into the wood, wherein it can conceal itself, and at the same time feast upon the fragments, as is proved by the presence of woody dust within its interior. Its tunnels are made in an oblique direction, not very deeply sunk below the surface, so that after a while the action of the waves washes away the thin shell, and leaves a number of grooves on the surface. Below these, again, the creature bores a fresh set of tunnels, which in their turn are washed away, so that the timber is soon destroyed in successive grooved flakes.

According to Mr. Allman, its habits can be very easily watched, as if it is merely placed in a tumbler of sea water, together with a piece of wood, it will forthwith proceed to work, and gnaw its way into the wood. The apparatus with which it works this destruction is a kind of file or rasp, which reduces the wood into minute fragments. In this creature the jaw feet are furnished with imperfect claws, and the tenth segment from the head is curiously prolonged into a large and long spine. The great flattened appendages near the tail seem to be merely used for the purpose of cleaning its burrow of wood dust which is not required for food. The creature always swims on its back, and when commencing its work of destruction, clings to the wood with the legs that proceed from the thorax. The wood-boring shrimp is one of the jumpers, and, like the sand hopper, can leap to a considerable height when placed on dry land. It has been detected in timber taken from the sea at Trieste. It was first observed as an inhabitant of the British seas several years ago, by Mr. Robert Ball, of Dublin, and in January, 1847, it was described by Mr. Mullins, C.E., in a paper read before the Institution of Civil Engineers of Ireland, as being very injurious to the timber piles in Kingstown Harbour, near Dublin, and far more destructive than the *Limnoria terebrans.*

We have already referred to the *lesson* the celebrated engineer, Brunel, received from observing the *teredo;* and we can state that architects have also received *lessons*

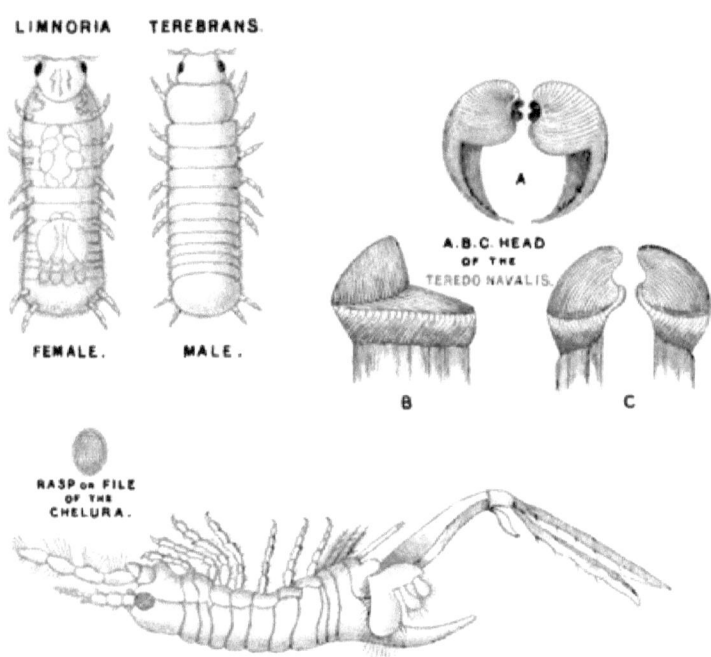

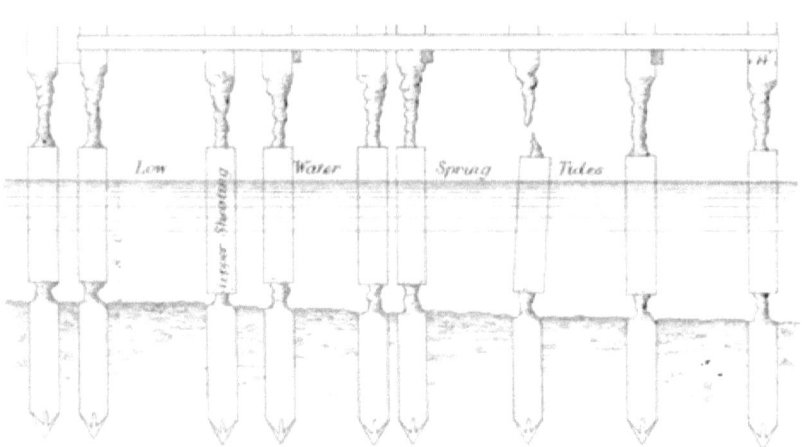

ELEVATION OF PILES, SOUTHEND PIER, DESTROYED BY THE TEREDO AND LIMNORIA ABOVE AND BELOW THE COPPER SHEATING.

from nature. Sir Christopher Wren constructed his spire of St. Bride's Church, London, after observing the construction of the delicate shell, called *Turretella*, which has a central column, or newel, round which the spiral turns. Brunelleschi designed the dome of Sta. Maria, at Florence, after studying the bones of birds and the human form; and Michael Angelo followed Brunelleschi in constructing the dome of St. Peter's, Rome.*

The LEPISMA is also a destructive little animal, which begins to prey on wood in the East Indies, as soon as it is immersed in sea water. The unprotected bottom of a boat has been known to be eaten through by it in three or four weeks.

These worms, it must be remembered, do not live except where they have the action of the water almost every tide, nor do they live in the parts covered with sand. The wooden piles of embankments and sea locks suffer very much from their depredations, and in the sea dykes of Holland they cause very expensive annual repairs.

The Dutch used to coat their piles with a mixture of pitch and tar, and then strew small pieces of cockle and other shells, beaten almost to powder, and mixed with sea sand, which incrusted and armed the piles against the attacks of the *teredo*. We believe it was a frequent practice in London, about half a century ago, to place small shells in the wooden pugging between the floor joists to deaden sound.

Having described the chief peculiarities of these worms,

* Note geometrical framing in spider's web.

shown their mode of working, and the extent to which their destructive powers may be carried, it will now be necessary to consider the various schemes which have been proposed and tried to prevent their desolating ravages. These may be divided into *three* classes, viz. the natural, chemical, and mechanical.

1st. By using woods which are able to resist the attacks of sea worms.

2nd. By subjecting piles to a chemical process.

3rd. By adopting a mechanical process.

First. We have not any English woods which resist their attacks. Elm (used for piles in England) or beech (used for piles, if entirely under water, in France) cannot withstand the *teredo*; while oak cannot battle successfully against wood-beetles in carvings. It is therefore necessary to inquire whether foreign woods are any better.[*]
Unfortunately the great expense of importing them into England prevents their use for piles.

Nearly all our foreign woods used for engineering and building purposes come from the Baltic or Canada: they are fir and pine. Memel timber from the Baltic is comparatively useless unless thoroughly creosoted; and

---

[*] 'Reports of the Juries,' Exhibition, 1851. 'Reports' by Dr. Gibson, Conservator of Forests, Bombay Presidency. 'Reports' by Dr. Cleghorn, Conservator of Forests, Madras Presidency. 'Reports' by Mr. H. B. Baden Powell, Inspector-General of the Forest Department, India, 1875. 'Reports' on the Teak Forests of Tenasserim, Calcutta, 1852. Papers by Mr. Mann and Mr. Heath on 'Decay of Woods in Tropical Climates,' Inst. C.E., 1866. Paper on 'The Ravages of the Limnoria Terebrans,' by Mr. R. Stevenson, Royal Society, 1862. 'Account of the Bell Rock Lighthouse,' by Robert Stevenson, 1824. Stevenson's 'Design and Construction of Harbours.' Smeaton's 'Reports.'

Canadian timber is not so good as the Baltic wood. At Liverpool and some of the western ports of England Canadian timber is preferred to Baltic, although we believe the reason to be that they cannot get the latter, except in small quantities at a time.

The following is a list of timber woods which, according to good authorities, resist for a long period of time the attacks of sea worms. It should be borne in mind, however, that the timber should be cut, during the proper season, from a large and full-grown tree; and, to prevent splitting, it should be kept from the direct action of the sun when first cut; it should have all the bark and sapwood removed, and allowed to dry a certain time before being used.

WOODS WHICH RESIST SEA WORMS.

Australia, Western.—Jarrah, beef-wood, tuart.
Bahama.—Stopper-wood.
Brazil.—Sicupira, greenheart.
British Guiana.—Cabacalli, greenheart, kakarilly, silverballi (yellow).
Ceylon.—Halmalille, palmyra, theet-kha, neem.
Demerara.—Bullet, greenheart (purple heartwood), sabicu.
India.—Malabar teak, sissoo, morung sál, dabu, than-kya, ilupé, anan, angeli, may-tobek. (Teak resists the *teredo*, but is not proof against barnacles.)
Jamaica.—Greenheart.
North America.—Locust.
Sierra Leone.—African oak, or tortosa.
South America.—Santa Maria wood.
Philippine Islands.—Malacintud, barnabá, palma-brava.
Tasmania.—Blue gum.
West Indies.—Lignum vitæ.

Second. The chemical, viz. Kyan's process of corrosive sublimate; Payne's process of sulphate of iron and muriate of lime; pitching and tarring; Burnett's process of chloride of zinc; and arsenic, or other mercurial preparations, have all failed, with the exception of Bethell's process of oil of tar. The failure must proceed from one of two causes; either that the sea-water decomposes the poisonous ingredients contained in the wood, or that these poisonous compounds have no injurious effect on the worms; it appears, however, that both these causes have been in operation, principally the latter.

Without a series of the most minute experiments, it is impossible to form any general notion of the action of sea-water on timber. Common salt, chlorides of calcium and magnesium, sulphate of soda, iodides and bromides of the same metals, are known to exist in sea-water, and in great abundance in the torrid zone. What effect these different ingredients may have upon saturated timber it is difficult to say, but it is extremely probable that they do have an effect.

With regard to the different poisonous compounds having no injurious effect on the worms, it should be remembered that all cold-blooded animals are much more tenacious of life than those of a higher temperament, and in descending the scale of animal creation, the tenacity of life increases, and this principle is more developed. A frog, which though cold-blooded, is an animal of a much higher order than the *teredo*, will not only live in hydrogen gas, but also in a strong solution of hydrocyanic acid, while at the same time a single drop placed on the nose of a rat, or in the eye of a rabbit, would produce instant death. A somewhat similar occurrence is noticed in the

'British and Foreign Medical Review,' for July, 1841, showing the slow effects of prussic acid on the common snake and turtle.

It may therefore be inferred, that as it requires a large quantity of the most virulently poisoned matter to destroy animals of a much higher order than the *teredo*, it would take a still greater quantity to affect those animals as they exist in their own element.

The preserving property of soluble salts, such as corrosive sublimate, sulphate of copper, &c., was considered to be founded upon their power of coagulating the albumen, and the sap of wood, thereby rendering that sap less liable to decay; but that very quality of combining with the albumen, destroyed the activity of the poison of the salts. A given quantity of corrosive sublimate of mercury, which if administered to a dog would kill it, would, when mixed with the white of an egg, become coagulated, and if swallowed in that state would be perfectly harmless; so a piece of wood, saturated by those salts, could be eaten by a worm without injury.

A French naturalist, M. de Quatrefages,* in 1848, suggested that a weak solution of mercury (corrosive sublimate) thrown into the water will destroy the milt of the *teredo*, and consequently prevent fecundation of the eggs, thus exhausting the molluscs in the bud. He proposed that ships should be cleared of this terrible pest by being taken into a closed dock, into which a few handfuls of corrosive sublimate should be thrown and

* See 'Sur un Moyen de Mettre tous les Approvisionnements de Bois de la Marine de la Piqûre des Tarets' (Compte. rend., Janv. 1848).

well mixed with the water. He considered that about 1 lb. of sublimate would be sufficient for 20,000 cubic metres (metre = 39·37 English inches) of water; but on account of the cost it would be advisable to use salts of lead or copper. This proposition of de Quatrefages reminds us of Chapman's suggestion, in 1812, to get rid of dry rot in ships, viz. by sweeping out the hold, laying from two to four tons of copperas in her bottom, and as much fresh water let in upon it as would make a saturated solution to soak into the wood.

M. de Quatrefages placed the four salts he used in his experiments in the following order, according to merit: 1st, corrosive sublimate; 2nd, acetate of lead; 3rd, sulphate of copper; and 4th, nitrate of copper.

In America, white oxide of zinc is used as a marine paint for ships and piles. In the United States Navy Yard at Gosport it is spoken well of, and very frequently employed. It is said to be much superior to white-lead, red-lead, verdigris, or coal-tar, and that timber covered with two coats of white zinc is neither attacked by the worm, nor do barnacles attach to it when immersed in salt water.

We can only find one instance of timber impregnated with water-glass having been tested against this subtle foe. Water-glass is certainly worth a further trial.

The instance we refer to occurred about forty years ago. In 1832, Dr. Lewis Feuchtwanger, of New York, was permitted by the Ordnance Department, under the direction of Commodore Perry, to perform experiments with water-glass on piles in the Brooklyn Navy Yard, and in various docks. The piles in the docks were destroyed by the *teredo* so fast that they had to be replaced every three years. The experiments proved highly satisfactory: the piles which had been so treated lasting many years, without any indication of being attacked by sea-worms.

The reader is referred to some works on water-glass mentioned below,* which are worthy of attentive perusal.

Third. The mechanical processes. They are few in number, and rather expensive.

At Saint Sebastian, in Spain, the piles of the wooden bridge standing in the sea have been guarded against the attacks of sea-worms in the following manner: Each pile is surrounded by a wooden box, and the space between filled up with cement. After six years it was proved that the piles were in a perfect condition, whilst the outer boxes were completely riddled by the worms. A similar method to this was adopted, some years since, to many of the piles in the Herne Bay Pier, which were affected by sea-worms. Several attempts had been made to protect the timber, by saturating it under various processes, with, however, only doubtful success. At last, a wooden casing was formed round each pile, leaving a space of about an inch all round, which was rammed full of lime or cement concrete. That process appeared to be perfectly successful, as the pier-master, who first adopted the method, stated that some of the piles had been so treated for three or four years, and although the worms had commenced

* 'Report of German Commission relative to rendering Woodwork and Stage Materials Incombustible.' Professor Fuch's and Dr. Pettenkofer's Reports. Dr. Feuchtwanger's works. M. Kuhlman's pamphlet. 'Reports relative to Ransome's Process.' Note M. Szerelmey's patent, 21 July, 1868.

their ravages, they appeared to have been checked, and not to have been able to exist when so enclosed.

In 1835, Brunel suggested an easy way of defending piles, which was to give them in the first instance a coat of tar; then powder them with brick-dust, which would render the wood sufficiently hard to receive a coat or two of cement. This is similar to the Dutch method.

Some foreigners use sheet lead nailed on to piles, and wrapped close round with well-tarred rope.

Copper sheathing has often been used for the protection of piling in piers and harbours. The destruction of copper by the action of sea-water is a matter which has long occupied the attention of scientific men, and it appears to be well ascertained that the decay does not result from the bad quality of the copper, for, according to Mr. Wilkinson, no difference could be discovered between the composition of copper that had endured well, and that which had been rapidly destroyed. Copper sheathing was used at Southend, but without success, for although nearly all the piles were covered with it for about 9 feet or 10 feet, the *limnoria* not only penetrated between the copper and the timber, but the copper had decayed to such an extent as in some cases to be no thicker than the thinnest paper; it was soft, and peeled off the wood very easily, and in two or three years would probably have been entirely destroyed.

Covering the surface of the timber with broad-headed scupper nails, arranged in regular rows with their heads at no great distance from each other, is a method which has been satisfactorily employed in various parts of the world, in Swedish and Danish vessels, even up to the present time, and, indeed, it was also practised by the Romans. The scupper-nailed piles at Southend, after twelve years' exposure to the sea, were perfectly sound, and although the nails were not driven close together in the first instance, yet the corrosive action was so great as to form a solid impenetrable metallic substance, upon which the worms refused to settle. Scupper nails have been proved at Yarmouth, as well as at other places, to have protected timber for forty years, but the process is expensive, as it costs one shilling per square foot. They should be about half an inch square at the head.

Captain Sir Samuel Brown, R.N., states that from numerous experiments and observations, he is satisfied that at present there is really no specific remedy against the attacks of sea-worms upon timber, except iron nails. He proposes to encase the piles with broad-headed iron nails resembling scupper nails, but considerably larger, and he says that in the course of a few months corrosion takes place, and spreads into the interstices. The rust hardens upon the pile, and becomes a solid mass which the worm will not touch. Experiments tried at the Trinity Pier, Newhaven, and Brighton Pier, have established the effectiveness of his method.

At the Cape of Good Hope, and many other places, wood piles are cased in iron, and occasionally iron piles are used instead of wood, at great cost. Further experience is desirable as to the durability of cast iron * in salt

---

* See 'Memoirs on the Use of Cast Iron in Piling,' by Mr. M. A. Borthwick, 'Trans. Inst. Civ. Eng.,' vol. i. No. 22.

water, especially as to its peculiar property of conversion, after a few years' immersion in the sea, into a carburet of iron, closely resembling plumbago, so that it may be easily cut with a knife. This, of course, diminishes its powers of resistance acting upon the framing it is intended to strengthen. In the course of the construction of the Britannia Bridge, about one hundred thin plates were delivered, which were not used on account of some error in their dimensions. They were left on the platform alongside the straits, exposed to the wash and spray of the sea; and after about two years were literally so completely decomposed as to be swept away with a broom into the water, not a particle of iron remaining.

We have already stated that the chemical processes have failed with the exception of Bethell's process of oil of tar, generally known as the creosoting process. This method, *when properly carried out*, thoroughly protects wood from the ravages of the *teredo* and other marine worms. The breakwaters and piers at Leith, Holyhead, Portland, Lowestoft, Great Grimsby, Plymouth, Wisbeach, Southampton, &c., have been built with creosoted timber, and in no case have the *Teredo navalis, Limnoria terebrans,* or any other marine worms or insects been found to attack these works, as certified to by the engineers in whose charge the several works are placed. In the cases of Lowestoft and Southampton we are enabled to give the detailed reports.

A most searching examination, lasting many days, was made in 1849, upon every pile in Lowestoft Harbour, by direction of Mr. Bidder; and the report of Mr. Makinson, the Superintendent of Lowestoft Harbour Works, contains the subjoined statement:

" The following is the result, after a close and minute investigation of all the piles in the North and South Piers.

" *North Pier.*—The whole of the creosoted piles in the North Pier, both seaward and inside the harbour, nine hundred in number, are sound, and quite free from *teredo* and *limnoria*.

" *South Pier.*—The whole of the creosoted piles in the South Pier, both seaward and inside the harbour, seven hundred in number, are sound, and quite free from *teredo* and *limnoria*.

" There is no instance whatever of an uncreosoted pile being sound. They are all attacked, both by the *limnoria* and the *teredo*, to a very great extent, and the piles in some instances are eaten through. All the creosoted piles are quite sound, being neither touched by the *teredo* or the *limnoria*, though covered with vegetation, which generally attracts the *teredo*."

There was only one instance of a piece of creosoted wood, in Lowestoft Harbour, being touched by a worm, and that was occasioned by the workmen having cut away a great part of one of the cross heads, leaving exposed the interior or heart of the wood, to which the creosote had not penetrated. At this spot a worm entered, and bored to the right, where it found creosote; on turning back and boring to the left, but finding creosote all around, its progress was stopped, and it then appeared to have left the piece of wood altogether.

In 1849, Mr. Doswell, who had the conduct of experiments on different descriptions of wood at Southampton, where the river was so full of the worm that piles of 14 inches square had been eaten down to 4 inches in four years, reported as follows: "From my examination, last spring tides, of the specimen blocks attached, on the 22nd February, 1848, to some worm-eaten piles of the Royal Pier, I am enabled to report that Bethell's creosoted timbers all continue to be unaffected by the worms; that the pieces saturated with Payne's solution continue to lose in substance by their ravages; and that the unprepared timbers diminish very fast, except the American elm, which stands as well (or nearly so) as that prepared by 'Payne's solution.'"

The following are the detailed particulars:

BETHELL'S CREOSOTED BLOCKS, PLACED FEBRUARY 22, 1848.

Memel, at low water of spring tides	} Unaffected by worms.
Red pine, at low water of neap tides	
Yellow fir, at high water of neap tides	A few barnacles.

PAYNIZED BLOCKS, PLACED APRIL 6, 1848.

Red pine, at low water of spring tides	Worm-eaten.
American elm, at low water of neap tides	} A few barnacles.
Fir, at high water of neap tides	

UNPREPARED BLOCKS, PLACED APRIL 6, 1848.

Memel, at low water of spring tides	Much worm-eaten.
American elm, at low water of neap tides	A few barnacles.
Fir, at high water of neap tides	Much worm-eaten.

On 1st January, 1852, Mr. Doswell ascertained that, notwithstanding the number of *teredines* and *limnoria* to be found in the Southampton Waters, none of the creosoted blocks had been attacked by them.

According to M. Forestier, similar results have been obtained at Brighton, Sunderland, and Teignmouth.

Allusion has already been made to Mr. Pritchard, of Shoreham, with reference to preserving timber. On July 26, 1842, he presented a report to the Treasurer of the Brighton Suspension Chain Pier Company, upon the preservation of timber from the action of sea-worms. We give a portion of it, as follows:

"Stockholm tar has been used, and proved to be of little service; this tar is objectionable owing to its high price, and also from its being manufactured from vegetable substances. All tars containing vegetable productions must be detrimental to the preservation of timber, especially when used in, and exposed to, salt water. This tar does not penetrate into the wood, and in a very few months the salt acid of the sea will eat it all away.

"Common gas or coal tar has been used to a great extent, and its effects are apparent to all. It does a very great deal of harm, forms a hard or brittle crust or coat on the wood, and completely excludes the damp and unnatural heat from the possibility of escape, owing to its containing ammonia, which burns the timber, and in a few years it turns brown and crumbles into dust. Indeed, timber prepared with this tar will be completely destroyed on this coast and pier by the ravages of the *Teredo navalis*, and the *Limnoria terebrans*, in five or six years.

"Also Kyan's patent, or the bi-chloride of mercury, has been used, but has proved equally useless. The sleepers Kyanized five years ago, and in use at the West India Dock warehouses, have been discovered to decay rapidly, and the wooden tanks at the Anti-Dry-Rot Company's principal yard are destroyed.

"I would recommend you for the future to use 'oil of tar and pyrolignite of iron' (Bethell's patent). This process will, without a doubt, succeed. I have proved in hydraulic works on this coast that it will fully prevent the decay in timber piles, destroy sea-worms, and supersede the necessity of coating the piles with iron nails. In Shoreham Harbour, for instance, there is a piece of red pine accidentally infused with pyrolignite of iron, which after being in use twelve years is perfectly sound. There is another waleing piece, the very heart of English oak, Kyanized, and in use only four years, which is like a honeycomb or network, completely eaten away by the *teredo* and other sea-worms. I have fully proved the efficiency of this method at different harbours and docks. Sixteen years ago I had timber prepared with it, and in use on the shores of the Dee, and it is at the present moment perfectly sound. The pyrolignite of iron must be used of very pure quality; the timber must be dry; afterwards the oil of tar must be applied, and not on any account must it contain a particle of ammonia. The immense destruction on the coast of timber by the sea-worms, and the important fact that at the Chain Pier there are not twenty of the original piles remaining at the present time, is of itself sufficient to awaken anxiety."

With regard to the opinion of foreigners on the subject of creosoting, we cannot do better than quote the report of the commission or committee (instituted in 1859) of the Royal Academy of Sciences, Holland, upon the means of preserving wood from the *teredo*, published at Haarlem in 1866. It is as follows:

"To conclude, it results from experiments which the committee has directed during six consecutive years, that—

"1st. Coatings of any sort whatever applied to the surface of the timber in order to cover it with an envelop upon which the young *teredo* will not fasten offer a very insufficient protection; such an envelop soon becomes damaged, either by mechanical action, such as the friction of water or ice, or by the dissolving action of water; and as soon as any point upon the surface of the wood is uncovered, however small it be, the *teredoes* of microscopic size penetrate into the interior of the wood.

"Covering wood with plates of copper, or zinc, or flat-headed nails are expensive processes, and only defend the wood as long as they present a perfect and unbroken surface.

"2nd. Impregnation with soluble metallic salts generally considered poisonous to animals does not preserve the wood from the invasions of the *teredo;* the failure of these salts is partly attributable to their being soaked out of the wood by the dissolving action of the sea-water, partly also to the fact that some of these salts do not appear to be poisonous to the *teredo*.

"3rd. Although we cannot venture to say that there may not be found in the colonies a wood that may resist the *teredo*, yet we may affirm that hardness of any timber is not an obstacle to the perforations of this mollusc. This has been proved by the ravages it has made on the Gaïac and Mamberklak woods.

"4th. The only means which can be confidently regarded as a preservative against the ravages of the *teredo* is the creosote oil; nevertheless, in the employment of this agent great care should be taken regarding the quality of the oil, the degree of penetration, and the quality of the wood treated."

These results of the experiments of the committee are confirmed by the experience of a large number of engineers of ponts et chaussées (bridges and causeways) in Holland, England, France, and Belgium. For example, very lately a Belgian engineer, M. Crépin, expressed himself as follows in his Report, dated 5th February, 1864, upon experiments made at Ostend:

"The experiment now appears to us decisive, and we think we may conclude that fir timber well prepared with creosote oil of good quality is proof against the *teredo*, and certain to last for a long time. Everything depends, therefore, upon a good preparation with good creosote oil, and on the use of wood capable of injection. It appears that resinous wood is easiest to impregnate, and that white fir should be rejected."

M. Forestier, the able French engineer at Napoléon-Vendée, sums up as follows the results of the experiments undertaken by him in the port of Sables-d'Olonne, viz.:

"These results fully confirm those obtained at Ostend, and it appears to us difficult not to admit that the experiments of Ostend and Sables d'Olonne are decisive, and prove in an incontestable manner that the *teredo* cannot attack wood properly creosoted."

It thus appears that there are three preservative

methods, which, according to experience, will **save timber piles from the ravages of the worms,** viz.: 1st. By using woods able to resist unaided their attacks. 2nd. The mechanical method, which is, **by covering the piles with scupper nails, &c.** This process is, however, very expensive, **especially** as the four sides of the pile must be **covered;** and, moreover, it affords **no protection to the timber from internal rot or decay.** 3rd. The chemical, or "creosoting" method. This process is cheaper than the last; it preserves the wood from decay, and no worms will touch it.

When unprepared piles are placed in the sea, there is every probability, sooner or later, of their being attacked by the *teredo*. This animal, however, is not left in peaceable enjoyment of the dwelling which it has constructed, and the food which it loves, but is liable to be attacked by an enemy, an *annelide*, to which the late M. de Haan has given the name of *Lycoris fucata*. This animal is to be found wherever the *teredo* exists, indeed its eggs and larva are to be met with in the midst of those of the mollusc. M. Kater has remarked that the adult *lycoris* dwelling in the mud which it enters during winter, and into which the piles are driven, climbs up the pile to the hole formed by the *teredo*, where, in some manner, it sucks or eats its victim; then having enlarged the entrance to the hole, it enters and rests in the place of the *teredo*. After a time it goes back to the entrance, and commences to seek for fresh prey.

The *lycoris* is narrow and not very long, provided laterally with a great many little feet terminating in points and covered with hair, and having in front a pair of hard superior jaws, pointed horns, and the inferior jaws bent round in the form of hooks. Behind the head are four pairs of tubuliform gills. It is with these arms that this little animal pursues and devours the *teredo*.

One day M. Kater was fortunately able to observe the operations of the *lycoris*. One of these animals coming out of a hole in the wood which he inhabited, seized upon a *teredo*, which M. Kater had previously deposited at the bottom of the vessel containing the wood. He saw the *annelide* seize the *teredo*, hurry away with it to the hole which he occupied, and so completely devour it that he finally left only the two valves of the shell. Our illustrations of the *teredo* and *lycoris* are derived from the works of Mr. Paton and M. Forestier; and our own sketches.

If the *lycoris* would only destroy the *teredo*, when the mollusc was in its infancy, what an invaluable little annelide it would be!

It appears to us a great pity that the woods we have named, or some of them, are not brought over to England in large quantities for harbour works. In Ceylon and India, the trees are felled by Indian wood-cutters at little cost; they are then dragged to the river banks by elephants or buffaloes, to be floated down the rivers to the different ports, so that labour is cheap. The question then remains, how to get the woods to England? When the 'Great Eastern' ship has finished carrying cables, perhaps its owners will not object to send the ship on a few voyages with heavy cargoes to India, Demerara, &c., bringing home " teredo-proof woods," *at moderate charges for freight?*

Finally, to place the subject in a practical form, we think the Institute of Civil Engineers, of London, would be heartily thanked by the engineering world if they would appoint a committee to inquire into the damages done to works by sea-worms; why they are found in some parts of a roadstead or harbour, and not in others; to consider the different remedies which have been proposed, their cost, and method of application; what course should be adopted to prevent sea-water injuriously affecting iron piles; and lastly, to publish a detailed account of their experiments and recommendations.

CHAPTER VIII.

ON THE DESTRUCTION OF WOODWORK IN HOT CLIMATES BY THE TERMITE OR WHITE ANT, WOOD-CUTTER, CARPENTER BEE, &c., AND THE MEANS OF PREVENTING THE SAME.

OF the ant proper, or that belonging to the order *Hymenoptera*, there are three species * in particular which attack timber, viz.:

1st. *Formica fuliginosa*, or black carpenter ant, which selects hard and tough woods.

2nd. *Formica fusca*, or dusky ant, which prefers soft woods.

3rd. *Formica flava*, or yellow ant, which also prefers soft woods.

The carpenter bee prefers particular kinds of wood. In India it is very fond of cadukai (*Tamil*) wood, which is often used for railway sleepers. Round the holes it makes there is a black tinge, arising, probably, from the iron in its saliva acting on the gallic acid of the timber. Providing it meets with the wood it prefers, it is not very particular whether it is standing timber, or the beams of a residence.

The termite, or white ant, is a terrible destroyer of wood in nearly all tropical countries. There are many species

---

* See Hurst's 'Tredgold's Carpentry,' p. 380, 1871. London.

of termite, and all are fearfully destructive, being indeed the greatest pest of the country wherein they reside. Nothing, unless cased in metal, can resist their jaws; and they have been known to destroy the whole woodwork of a house in a single season. They always work in darkness, and, at all expenditure of labour, keep themselves under cover, so that their destructive labours are often completed before the least intimation has been given. For example, the termites will bore through the boards of a floor, drive their tunnels up the legs of the tables or chairs, and consume everything but a mere shell no thicker than paper, and yet leave everything apparently in a perfect condition. Many a person has only learned the real state of his furniture by finding a chair crumble into dust as he sat upon it, or a whole staircase fall to pieces as soon as a foot was set upon it. In some cases the termite lines its galleries with clay, which soon becomes as hard as stone, and thereby produces very remarkable architectural changes. For example, it has been found that a row of wooden columns in front of a house have been converted into a substance as hard as stone by these insects. In pulling down the old cathedral *at Jamaica*, some of the timbers of the roof, which were of hard wood, were eaten away, and a cartload of nests formed by the ants was removed, after being cut away by great labour with hatchets.

The first indication of a house being attacked by ants in the tropics is, perhaps, the yielding of a floor board in the middle of a room, or the top hinge of a door suddenly leaving the frame to which it had been firmly screwed a short time before.

That the ants provide for winter—as not only Dr. Bancroft and many others, even King Solomon, reports—is found to be an error. Where there is an ordinary winter, the ants lie dormant, during which torpid state they do not want food.

The greater number of species belong to the tropical regions, where they are useful in destroying the fallen trees that are so plentiful in those latitudes, and which, unless speedily removed, might be injurious to the young saplings by which they are replaced. Two species, however, are known in Europe, namely, *Termes lucifugus* and *Termes rucifollis*, and have fully carried out their destructive character, the former species devouring oaks and firs, and the latter preferring olives and similar trees. *At La Rochelle* these insects have multiplied so greatly as to demand the public attention.

M. de Quatrefages, who visited one of the spots in which these destructive insects had settled themselves, gives the following account of their devastating energy: "The prefecture and a few neighbouring houses are the principal scene of the destructive ravages of the termites, but here they have taken complete possession of the premises. In the garden not a stake can be put into the ground, and not a plank can be left on the beds, without being attacked within twenty-four or forty-eight hours. The fences put round the young trees are gnawed from the bottom, while the trees themselves are gutted to the very branches.

"Within the building itself the apartments and offices are all alike invaded. I saw upon the roof of a bedroom that had been lately repaired galleries made by the termites which looked like stalactites, and which had begun to show themselves the very day after the workmen left the place. In the cellars I found similar galleries, which were either half way between the ceiling and the floor, or running along the walls and extending, no doubt, up to the very garrets, for on the principal staircase other galleries were observed, between the ground floor and the second floor, passing under the plaster wherever it was sufficiently thick for the purpose, and only coming to view at different points where the stones were on the surface, for, like other species, the termites of La Rochelle always work under cover wherever it is possible for them to do so. It is generally only by incessant vigilance that we can trace the course of their devastations and prevent their ravages.

"At the time of M. Audoin's visit a curious proof was accidentally obtained of the mischief which this insect silently accomplishes. One day it was discovered that the archives of the department were almost totally destroyed, and that without the slightest external trace of any damage. The termites had reached the boxes in which these documents were preserved by mining the wainscoting, and they had then leisurely set to work to devour these administrative records, carefully respecting the upper sheets and the margin of each leaf, so that a box which was only filled by a mass of rubbish seemed to enclose a file of papers in perfect order.

"The hardest woods are attacked in the same manner. I saw on one of the staircases an oak post, in which one of the clerks had buried his hand up to the wrist in grasping at it for support, as his foot accidentally slipped. The interior of the post was entirely formed of empty cells, the substance of which could be scraped away like dust, while the layer that had been left untouched by the termites was not thicker than a sheet of paper."

It is most probable that these insects had been imported from some vessel, as they attacked two opposite ends of the same town, the centre being untouched. M. de Quatrefages tried many experiments on these insects with a view of discovering some method of destroying them, and came to the conclusion that if *chlorine* could be injected in sufficient quantities, it would in time have the desired result.

The termite or white ant is represented by Linnæus as the greatest pest of both Indies, because of the havoc they make in all buildings of wood, in utensils, and in furniture. They frequently construct nests within the roofs and other parts of houses, which they destroy if not speedily extirpated. The larger species enter under the foundations of houses, making their way through the floors and up the posts of buildings, destroying all before them; and so little is seen of their operations that a well-painted building is sometimes found to be a mere shell, so thin that the woodwork may be punched through with the point of the finger.

Many kinds of wood *in Brazil** are impervious to the termite, which insect generally selects the more porous

---

* See Charlesworth's 'Magazine of Natural History,' 1838, Art. *Myrmica domestica*. Also, 'Boston Journal of Natural History,' 1834, p. 993, Art. *Myrmica molesta*.

woods, and especially if these are in contact with the earth. In dry places, and with *a free circulation of air*, it does not prefer timber thus situated; and it is found that roofs of buildings of *good* and well-seasoned native wood resist for an indefinite period both the climate and the termite. As a general rule, Brazilian timber is very brittle.

It shows the difference of effects between one climate and another, that in Brazil the more porous and open-grained timbers are most subject to the attacks of the white ant, especially if they are in contact with the earth; but *in Australia* it is the reverse, for there it is the hardest description of timber that those insects first attack. There is one wood in particular, in common use, to which this remark applies, namely, "Iron Bark." Its density is so great that it sinks in water, and its strength is extraordinary, and yet the wood the white ants are particularly fond of. In the West Indies, the ants prefer hard woods.

At Bahia, the timber is less affected by the termite than *in Pernambuco;* but even in the latter place the white ant does not like dry places with a free circulation of air.

Mr. Shields, when on a short visit to Pernambuco, examined some timber bridges, and in one, which had only been constructed three years, he found the ends of the timber had been placed in contact with the moist clay; at those places he could readily knock off the crust of the wood, and the interior of the wood was almost filled with white ants: the decay was augmented by the contact of the wood with the moist clay. We have been informed that timber for the Government works is stored to the depth of about 1 foot 6 inches in the sea-sand, to protect it from the white ants and the *teredo*; and that in Pernambuco, since the establishment of the gas-works, the Brazilian engineers and constructors "pay" over the ends of all timbers used in buildings with coal-tar.

In Ceylon, no timber—except ebony and ironwood, which are too hard; palmyra, in *northern* Ceylon; and those which are strongly impregnated with camphor or aromatic oils, which they dislike—presents any obstacle to their ingress. Sir Emerson Tennant, in his work on Ceylon, says: "I have had a cask of wine filled, in the course of two days, with almost solid clay, and only discovered the presence of the white ants by the bursting of the corks. I have had a portmanteau in my tent so peopled with them in the course of a single night that the contents were found worthless in the morning. In an incredibly short time a detachment of these pests will destroy a press full of records, reducing the paper to fragments; and a shelf of books will be tunnelled into a gallery, if it happened to be in their line of march."

In Ceylon, the huts are plastered over with earth, which has been thrown up by white ants, after being mixed with a powerful binding substance (produced by the ants themselves), and through which the rain and moisture cannot penetrate. This will hold the walls together when the entire framework and the wattles have been eaten, or have become decayed.

In the Philippine Islands, ambogues, a strong, durable wood, suffers much from the termites. Sir John Bowring, in his work on these islands, thus writes of the ravages of the white ants in the town of Obando, Province of Bulacan, Philippine Islands: "It appears that on the 18th March, 1838, the various objects destined for the services of the mass, such as robes, albs, amices, the garments of the priests, &c., were examined, and placed in a trunk made of the wood called 'narra' (*Pterocarpus palidus*). On the 19th they were used in the divine services, and in the evening were restored to the box. On the 20th some dirt was observed near it, and, on opening it, every fragment of the vestments and ornaments of every sort were found to have been reduced to dust, except the gold and silver lace, which were tarnished with a filthy deposit. On a thorough examination not an ant was found in any other part of the church, nor any vestige of the presence of these voracious destroyers; but five days afterwards they were discovered to have penetrated through a beam 6 inches thick."

The red ant *in Batavia* (north-west end of Java) is another devastator. The red ant contains formic acid (acid of ants) and a peculiar resinous oil. Thunberg* has found cajeput effectual in destroying the red ants of Batavia: he used it to preserve his boxes of specimens from them. When ants were placed in a box anointed with this oil, they died in a few minutes.

In Surinam, Guiana, several species of worms are produced in the palm-trees as soon as they commence to rot: they are called "groo-groo," and are produced from the spawn of a black beetle; they are very fat, and grow to

* 'Thunberg's Travels,' vol. ii. p. 300.

the size of a man's thumb. The groo-groo will very quickly destroy wood which has commenced to rot.

In Surinam, Captain Stedman* was obliged to drive nails into the ceiling of his room, and hang his provisions from the nails; he then made a ring of dry chalk around them, very thick, which crumbled down the moment the ants attempted to pass it. In Guiana, the young ants will swim across a small pool of water to get at sugar; some get drowned, the rest get the sugar.

In Japan, according to Kœmpfer,† ants do considerable damage to wood.

In Senegal, the ant (*Termite belliqueux*) is a formidable agent of destruction. In a season, all the carpentry of a house is destroyed by them. Spartmann, in his 'Voyage to the Cape of Good Hope,'‡ gives an excellent account of their methods of working.

The *Termite lucifuge* has been discovered in the environs of *Bordeaux*, in the pine-trees; also in the marine workshops at *Rochefort*. It is believed to have been imported from America.

The *Termite flavicole*, a few years since, attacked the olive-trees of *Spain*, and it occasionally visits the centre of *France*.

White or yellow pine wood can only be used in the tropics for doors, movable window frames, bodies of railway waggons, or other work intended to be kept in motion. Its use even for these purposes is questionable, as the

---

* 'Expedition to Surinam.' By Captain Stedman. 1813. London.

† Kœmpfer's 'Japan,' vol. ii.

‡ 'Voyage de Spartmann au cap de Bonne-Espérance: voy. *Dict. d'Hist. Nat.* de Guérin.' 1839.

white ant has such an affinity for it, that a door or a window which has remained shut for a few weeks will almost invariably be attacked by that insect.

North American pitch pine withstands very well the attacks of the termite, when used in the roofs of buildings, or in any locality not humid; but it is found after a time, when laid upon the earth, to lose its resisting powers, as well as to become subject to rapid decay.

"Greenheart" timber in its natural state is proof against the attacks of this insect in tropical climates —especially that known as the "purple-heart" wood. There are two reasons why it enjoys this immunity from attack: first, there is its great hardness; and, secondly, there is the presence of a large quantity of essential oil. It is very hard and durable wood; a little heavier than water. It is obtained at Demerara.* Great care is required in working it, as it is very liable to split. In sawing it is necessary to have all the logs bound tightly with chains, failing which precaution the log would break up into splinters, and be very apt to injure the men working it.

"Jarrah" wood, from Australia, is also proof against the attacks of the white ant. It is occasionally liable to shakes.

"Panao" wood, from the Philippine Islands, gives the talay oil, which destroys insects in wood.

"Bilian" wood is imported to Bombay, from Sarawak,

* See Paper by Mr. J. B. Hartley, read before the Institution of Civil Engineers, London, 23rd June, 1840, "On the Effects of the Worm on Kyanized Timber exposed to the Action of Sea Water; and on the Use of Greenheart Timber from Demerara."

Borneo. This wood is impervious to the attacks of the termite, and does not decay when under fresh or salt water, where it remains as hard as stone.

"Sál" wood, in India, is occasionally touched by the white ant. This wood, however, requires two years to season, and it will twist, shrink, and warp whenever the surface is removed, after many years' seasoning. Only about 2 lb. of creosote oil per cubic foot can be injected into sál wood. "Kara-mardá" is avoided by this little insect; but when used for planks it requires twelve to fifteen months' previous seasoning. "Neem-wood," used for making carved images, enables an image to remain undisturbed by the white ant.

The following is a list of woods which resist for a long time, if not altogether, the attacks of the termites, or white ants:

ANT-RESISTING WOODS.

America.—Butternut, pitch pine. (Pitch pine is sometimes attacked.)
Australia, Western.—Jarrah.
Borneo.—Bilian.
Brazil.—The sicupira assú, sicupira meirim, or verdadeiro, sicupira acari, oiticira, gararoba, paó saulo, sapucaia de Pilao, sapucarana, paó ferro, and imberiba, resist the white ant, *except* in the sapwood. The angelim amargozo, araroba, pitia, cocâo, bordâo de Velha, ameira de Sertao, parohiba, cedro, louro cheiroso, and louro ti, resist the white ant, *even* in the sapwood.
Ceylon.—Ebony, ironwood, palmyra, jack, gal-mendora, paloo, cohambe.
Demerara.—Greenheart.
Guiana, British.—Determa, cabacalli, kakatilly.
India.—Cedar, sál, neem, kara mardá, sandal, erul, nux vomica, thetgan, teak. (Ants will bore through teak to get at yellow pine.)
Indies, West.—Bullet wood, lignum vitæ, quassia wood.
Pernambuco (Brazil).—Maçaranduba (red), barubú (purple), mangabevia de Viado.
Philippine Islands.—Molave, panao.
Tasmania.—Huon pine.
Trinidad.—Sepe.

In piles of wooden sleepers which have been lying ready for use *in India* for about six months, at least 10 per cent. have been found destroyed by ants. It has been supposed that the jarring motion of a train on a railway would prevent the white ant from destroying the timber sleepers; but there is reason to doubt this, from the fact that on an examination of the 'Hindostan' steam vessel, a considerable portion of her timber framing was found to be eaten away by that destructive insect, particularly in the parts close to the engine and boilers, where there had been the greatest amount of vibration. The telegraph posts are particularly subject to their depredations so long as the timber is sunk in the ground; but when a metallic socket is supplied, the wood is safe from their visits. A further precaution is taken to preserve the lower end of the post by running liquid dammer into the metallic sheath, so that the enclosed part of the post is encased with a coating of resin. The telegraph wires when covered with gutta-percha (a vegetable substance) are also liable to their attacks.

Numerous expedients have been suggested for getting rid of this destructive insect, some of which have been successful, but the majority only partially so.

In India, the timbers of a house infested with white ants are periodically beaten to drive them away. Of course, this only succeeds for a short time, as they soon return.

The salt vessels plying on the coast of India use oil of tar, and a considerable quantity of castor-oil, mixed with cow-dung mortar, which, while it adheres to the wood, is an effectual protection against ants and rot. The earth oil, or Arracan oil, is considered as good as creosote to protect wood from ants. It can be obtained at Moulmein and Rangoon, in leathern bottles or skins, at about 6d. per gallon.

It used to be a practice *in the West Indies* to destroy whole colonies of ants which had built their nests either on trees or under the roofs of houses, by shooting powdered arsenic out of a quill into an orifice made into their covered ways, along which they ascended and descended from and to the ground.

It has been estimated that the depredations of the white ant in India costs the Indian Government 100,000*l.* a year, which is expended in repairing the woodwork of houses, barracks, bridges, &c.

When Dr. Boucherie gave up his sulphate of copper process for the use of the French public, he received *a national reward*. If the Indian Government is disposed to give us *a national reward*, we could show how it may save at least half the 100,000*l.* a year—which is expended in repairing the damages done by the white ants—with little trouble.

In the Madras Presidency periodical inspections have to be made, not only with regard to the white ant, but with respect to the presence and subsequent germination of vegetable matter or seeds in the mortar. In some instances, where proper precautions had not been taken, roots had formed very rapidly, and of such great size as to bodily dislodge by their pressure large stones from buildings.

Therefore, to prevent this germination, a proportion of "Jagherry," or coarse native sugar, varying from 2 per cent. in ordinary work, to from 5 per cent. to 8 per cent. in arch work, is mixed with the lime.

In 1856, in consequence of the ravages of the white ants in the King's Magazine, Fort William, India, the flooring and powder racks had to be reinstated. Captain A. Fraser, R.E., had the basement covered with concrete, 4 lb. of yellow arsenic being added to every 100 cubic feet of concrete. In the mortar used for the pillars arsenic was used in the proportion of ½ lb. to every 100 cubic feet of brickwork; a small quantity of arsenic was also mixed with the paint, and ½ lb. (four chittacks) of arsenic was also mixed with every 100 superficial feet of plaster. In 1859 the town mayor reported to the Government that no traces of white ants had since been found either inside or outside the building.

Colonel Scott, when Acting Chief Engineer, Madras Presidency, reported to the Government, December 24, 1858, that the following receipt was used for exterminating white ants in the Madras Presidency, and was found to be very successful:

	lb.	oz.	
Arsenic	2	4	
Aloes	2	4	
Chunam soap	2	13	(common country soap).
Dhobies mud	2	8	(Khar).

Pound the arsenic and aloes, scrape the soap, mix with mud, and boil in a large chatty half full of water until it bubbles; let it cool, and when cold, fill up with cold water. The mixture should boil for nearly an hour: it is applied as a wash.

The white ants of Calcutta are small in comparison with those of the upper provinces.

Colonel Scott, Chief Engineer at Bombay, records instances of timber being boiled under pressure in various antiseptic solutions, such as sulphate of copper, arsenious acid, and corrosive sublimate, with satisfactory results; but considerable apparatus is necessary, and the expense forbids its use except in large public works. On the other hand, in 1847, Mr. G. Jackson, being engaged under Mr. Rendel, C.E., on works in India, tried several experiments with Mr. J. Bourne, in order to test the possibility of preserving timber from the ravages of the white ant. Ninety pieces of wood, 9 inches long by 4 inches square, saturated according to the different processes of Burnet, Payne, and Margary,* under the direction of the patentees themselves, were experimented upon, in five situations, one with a considerable amount of moisture, and four dry; through inadvertency Mr. Bethell's specimens were only tested in the dry positions. The result was, that where there was moisture the timber was entirely destroyed, whilst where they were kept dry the result was better, but still not satisfactory. It seems difficult to account for these different results obtained by Colonel Scott and Mr. Jackson; but evidently the same strength of solutions, and the same qualities and descriptions of woods, cannot have been used by each gentleman.

Captain Mann and Captain McPherson painted the joists and planking of several buildings at Singapore with

---

* Margary's process failed to preserve wood from rot on the Bristol and Exeter Railway, England.

gambir composition, and **the result** was perfect **success,** although the buildings had been previously infested with white ants. **Gutta** gambir **is** juice extracted from the leaves of **a** plant of the same name (*Uncaria gambir*) growing in Sumatra, &c., inspissated by decoction, strained, suffered to **cool** and harden, and then **cut** into cakes **of** different **sizes, or** formed into balls. **Chief** places of manufacture, Siak, Malacca, and Bittany; gambir is now imported to England to **a** slight extent. **The** gambir composition referred to is made as follows: Dissolve three pints of gambir in twelve pints of dammer-oil over a slow fire; then stir one part of lime, sprinkling it over the top to prevent its coagulating and settling in a mass **at the** bottom; it must **be well** and **quickly** stirred. It should then be taken **out** of the cauldron, and **ground** down **like** paint on a muller till it is smooth, **and** afterwards returned to the pot and heated. **A little oil** should be added to make it tractable, and the composition can then be laid over the material. To be treated with a common brush. Against the *Teredo navalis* may be substituted the same proportion **of** black varnish **or** tar for dammer-oil, of course omitting the grinding down, which would not answer with tar.

Burnett's chloride of zinc process is said **to be a** good preservative for wood liable to be attacked by ants: the zinc penetrates to the heart of the **wood.**

Creosoted timber, **it is well** known, resists the attacks of the white ants; **but** the close grain of the generality of tropical timber renders any attempt to creosote it all but useless. **Of course,** creosoted fir timber could be, in fact is, exported from England, but the cost of freight and other charges will always make it very expensive, and be a great drawback to its general use abroad. Mr. J. C. Mellis, Engineer to the Government of St. Helena, writes in very high terms of creosoted timber as used there, where the white ant abounds. Between the years 1863 and 1866, experiments* were made with many specimens of woods (by order of the Lieutenant-Governor), in order to find out those which would resist the white ant. Teak remained uninjured; jarrah wood was partially destroyed; while pitch-pine, oak, cedar, ash, elm, birch, beech, and mahogany, were quite destroyed.

In Ceylon creosoted timber is not attacked by white ants, but the black coating, if exposed, renders it so heat-absorbing, that it is apt to split, and, unless thoroughly impregnated with the creosote, a road is opened to the inside, and the ants will soon destroy all that part which is unprotected.

Coal-tar will destroy white ants. Some years ago Mr. Shields took short baulks of timber where the ants had commenced operations, and tried the system of pouring a very small stream of coal-tar through the heart of the timber which the ants had hollowed out, and afterwards splitting it open to see the result. He found the white ants completely destroyed; they were shrivelled up like shreds of half-burnt paper by the mere effluvium of the coal-tar.

---

* See Paper by Mr. Thomas Hounslow, of the Royal Engineers' Department, published in 'Engineering,' p. 198, 21st September, 1866. Also, Hurst's edition of 'Tredgold's Carpentry,' page 380. 1871. London.

Creosoting is excellent for railway sleepers, piles, &c., but it will not do for buildings, which the white ants prefer. It is objectionable for dwellings; 1st, on account of its smell, which is disagreeable; 2nd, on account of its colour, black, which is unsightly; 3rd, on account of its inflammability.

With regard to the depredations of white ants, anything of a bitter taste injected into the fibre of the wood prevents their attacks, though it may not be so good as coal-tar; even a small quantity of turpentine has the effect of killing them instantly. Carbolic acid has been used, but its smell is objectionable. In South America, the leaves of the black walnut are soaked in water for some hours, then boiled; and when the liquid has cooled, it is applied to the skins of horses and other animals, to prevent their being bitten or "worried" by insects. We do not know if this has been used as a wash, or injected into wood, to prevent it being "worried" by ants.

It thus appears that there is no remedy generally adopted in tropical climates for preventing the depredations of the white ants; but there is one method very frequently adopted in hot countries of getting rid of them. It is a desperate remedy, we admit, but desperate cases frequently require desperate remedies: it is simply by EATING THEM. Europeans have pronounced the termites to be peculiarly delicate and well flavoured, something like sweetened cream. The termites are prepared for the table by various methods, some persons pounding them so as to form them into a kind of soft paste, while others roast them like coffee-beans or chestnuts. Termites, or

S

white ants, are eaten by various African tribes, both raw and boiled; and it is said that the Hottentots "get into good condition on this diet." In India, the natives capture great quantities of these insects, which they mix up with flour, producing a kind of pastry, which is purchased at a cheap rate by the poorer classes. In Ceylon, bears feed on the termites. Some of the Africans prepare large quantities of them for food, by parching them in kettles over a slow fire; in this condition they were eaten by handfuls as delicious food. The traveller Smeathman states that he often ate them dressed in this way, and found them to be "delicate, nourishing, and wholesome, resembling in flavour sugared cream, or sweet-almond paste." In Brazil, the yellow ants are eaten by many persons. Humboldt states that in some of the South American countries ants are mixed with resin and eaten as a sauce. In Siam, ants' eggs are considered a luxury; they are sent to the table curried, or rolled in green leaves mingled with fine slices or shreds of fat pork. In Sweden, ants are distilled along with rye to give a flavour to the inferior kinds of brandy. Chemists have ascertained that ants secrete a pleasant kind of vinegar, or a peculiar acid, called formic acid.

In Brazil, however, the eating process goes on extensively as follows:

1st. Ants eat the wood.
2nd. Ant-eaters eat the ants.
3rd. Woodsmen eat the ant-eaters.
4th. Wild animals eat the woodsmen.

Teak-oil, extracted from teak chips, was, in 1857, recommended by a Mr. Brown to the Government of St. Helena, through the Government of Madras. Timber coated with this oil, as reported to the Secretary to the Government of Madras by the several executive engineers of the Public Works Department, even when placed in a nest of white ants, was not touched by them. The cost of this oil, in certain experiments made by order of the Madras Government, in 1866, was reported to be $6\tfrac{3}{4}$ annas for $1\tfrac{1}{4}$ ounce, which is too expensive. In the central provinces the cost would be $1\tfrac{1}{4}$ anna per quart.*

In the East Indies there are several species of woodcutter (*Xylocopa*) and carpenter bee (*Xylocopa*), which confine their ravages to the wood after it has been felled. The wood-cutter tunnels through the beams and posts of buildings, which they frequent in great numbers. The passages are from 12 to 15 inches long, and more than half an inch in diameter. If the insects are numerous, their ravages are dangerously destructive, and they soon render beams unsafe for supporting the roof.

The carpenter bee of Southern Africa is one of those curious insects which construct a series of cells in wood. After completing their burrow, which is open at each end, they close the bottom with a flooring of agglutinated sawdust, formed of the morsels bitten off during the operation of burrowing, lay an egg upon this floor, insert a quantity of "bee-bread," made of the pollen of flowers and their juices, and then cover the whole with a layer of the same substance that was used for the floor. Upon this is laid another egg, another supply of bee-bread is inserted, and

* See Maconochie's suggestion, p. 163.

a fresh layer of sawdust superimposed. Each layer is therefore the floor of one cell and the ceiling of another, and the insect makes on the average about ten or twelve of these cells.

The carpenter bee destroys the woodwork of buildings in the north of Ceylon, but in the south of the island woodwork has two enemies to contend against, viz. the porcupine and a little beetle. The porcupine destroys many of the young palm-trees, and the ravages of the cocoa-nut beetle (*Longicornes*) are painfully familiar to the cocoa-nut planters. The species of beetle, called by the Singalese "cooroominya," is very destructive to timbers. It also makes its way into the stems of the younger trees, and after perforating them in all directions, it forms a cocoon of the gnawed wood and sawdust, in which it reposes during its sleep as a pupa, till the arrival of the period when it emerges as a perfect beetle. Mr. Capper relates that in passing through several cocoa-nut plantations, "varying in extent from twenty to fifty acres, and about two to three years old, and in these I did not discover a single young tree untouched by the cooroominya."

Sir E. Tennant thus writes of the operations of the carpenter bee on the wooden columns of the Colonial Secretary's official residence, at Kandy, Ceylon: "So soon as the day grew warm, these active creatures were at work perforating the wooden columns which supported the verandah. They poised themselves on their shining purple wings, as they made the first lodgment in the wood, enlivening the work with an uninterrupted hum of delight, which was audible to a considerable distance.

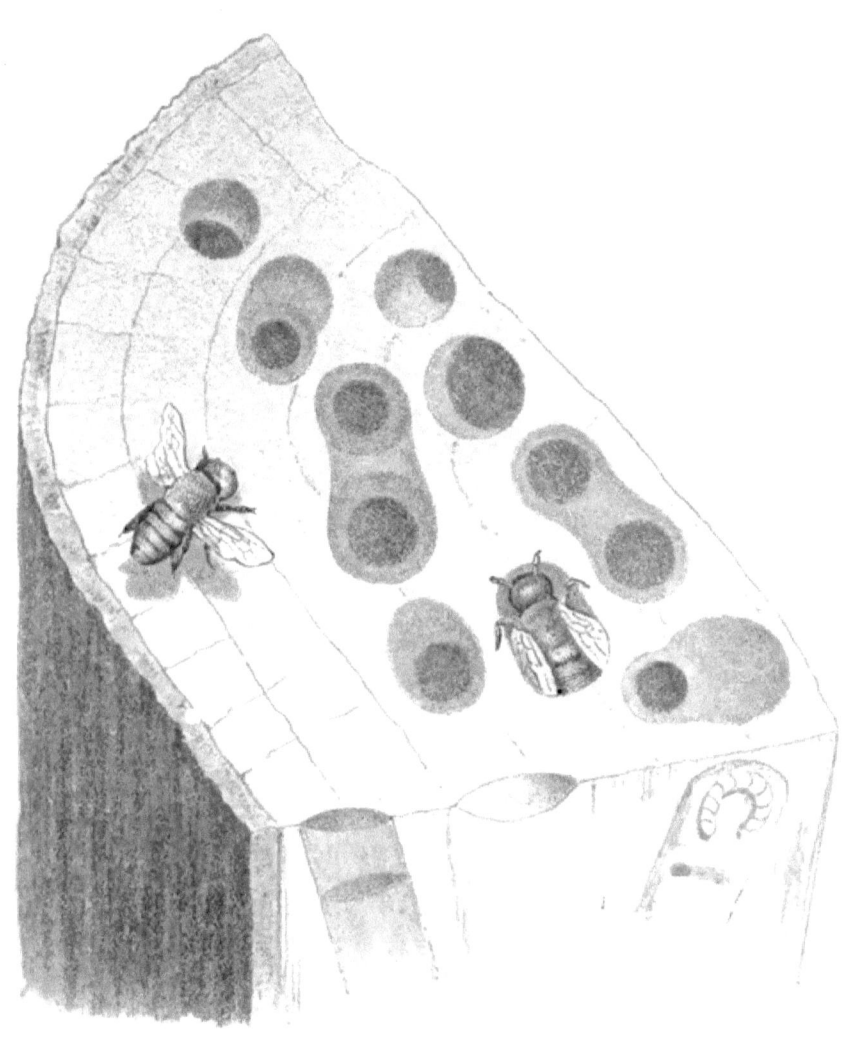

Carpenter Bees "at work".

When the excavation had proceeded so far that the insect could descend into it, the music was suspended, but renewed from time to time, as the little creature came to the orifice to throw out the chips, to rest, or to enjoy the fresh air. By degrees a mound of sawdust was formed at the base of the pillar, consisting of particles abraded by the mandibles of the bee; and these, when the hollow was completed to the depth of several inches, were partially replaced in the excavation, after being agglutinated to form partitions between the eggs, as they are deposited within."

Fortunately in England the owner of a house has no opportunity of watching ("with an uninterrupted hum of delight, audible to a considerable distance") the operations of the carpenter bee, on the wooden beams and posts of his building.

We must now *consider the ways* of the wood-beetle, which will be found described in the next chapter, and only write a few words before closing this. A modern engineer is no sluggard, of that we are certain; but if he intends erecting large buildings in any of the places abroad which we have referred to, he will find it very necessary to pay particular notice of the following words of King Solomon:

"Go to the ant, thou sluggard; *consider her ways,* and be wise."

Proverbs vi. 6.

CHAPTER IX.

ON THE CAUSES OF DECAY IN FURNITURE, WOOD CARVINGS, ETC., AND THE MEANS OF PREVENTING AND REMEDYING THE EFFECT OF SUCH DECAY.

ALTHOUGH trunks and boxes are of themselves of little importance, they derive great consequence from the valuable deposits of written papers, deeds, books, &c., which they frequently contain, that are subject to destruction from timber-destroying insects. It is well known that the smell of Russian leather, which arises from an essential oil, is a preservative of books. Leather or woods impregnated with petroleum, or with oil of coal-tar (which has a smell not much dissimilar) would be productive of the same effect, because known to be peculiarly obnoxious to insects: these oils are, however, very inflammable.

At all times beech-wood is exposed to the attacks of beetles, and it cannot be used, even for household furniture, without being impregnated with some kind of oil or varnish, as a defence against these insects—a very curious fact, for the growing trees are remarkably free from the attacks of wood-devouring insects. Larch being solid, and its juices hot, pungent, and bitter, is rarely affected with the larvæ of insects.

Mr. Westwood, Hope Professor of Zoology, Oxford, says: "The insects which in this country are found to be the most injurious from their habit of burrowing into the wood of furniture, belong to three species of beetles, of small size, and cylindrical in form (the better to enable them to work their way through the burrows in the wood), belonging to the family *Ptinidæ*, and known under the systematic names of *Ptilinus pectinicornis, Anobium striatum*, and *Anobium tessellatum*.

"In the perfect state, the insects of the genus *Anobium* are well known under the name of the "deathwatch," as these insects produce the ticking noise occasionally heard in old houses. It is also the *Anobium striatum* which is so injurious in libraries; the grub burrowing through entire volumes, and feeding upon the paper, and especially the *pasted* backs of the books.

"The destruction of these insects, when enclosed in articles of furniture, is by no means easy, although with care much mischief might be prevented. The saturation of the wood with some obnoxious fluid previous to its being used up in the manufacture of objects of value would be beneficial.

"A strong infusion of colocynth and quassia, spirits of turpentine, expressed juice of green walnuts, and pyroligneous acid, have all been proposed. In hot climates the ravages of the *Anobium* on books have been prevented by washing their backs with a fluid compound of corrosive sublimate (ten grains) and four ounces of alcohol, and the paste used in the book covers is there also mixed with alcohol."

Sir H. Davy and Professor Faraday hesitated to employ corrosive sublimate as a means of preventing the ravages of the bookworm in Earl Spencer's library, at Althorp, not feeling certain as to whether the quantity of mercury used would affect the health of the inhabitants. Amongst all the combinations of mercury, perhaps the bi-chloride, or corrosive sublimate, is the most terrible poison. It should be remembered that there are two chlorides of mercury—one the proto-chloride, ordinarily known as calomel; the other, bi-chloride, ordinarily known as corrosive sublimate; the respective compositions of which are as follows:

Calomel, and Corrosive Sublimate.	Parts by Weight.	
	Chlorine.	Mercury.
Calomel, or proto-chloride of mercury	36	200
Corrosive sublimate, or bi-chloride of mercury	72	200

Hence the ratio of chlorine in these two chlorides is as one to two.

Botanists have long used a solution of corrosive sublimate in alcohol, known by the name of Smith's solution, to preserve the specimens in their herbaria from the aggressions of insects.

The Rev. J. Wood, writes:[*] "I know to my cost sundry Kaffir articles being absolutely riddled with the burrows of these tiny beetles (*Anobium striatum*), and not to be handled without pouring out a shower of yellow dust, caused by the ravages of the larva. The most complete wreck which they made was that of a New Guinea bow, which was channelled from end to end by them, and in

[*] 'Insects Abroad.' By the Rev. J. Wood. 1874. London.

many places they had left scarcely anything but a very thin shell of wood.

"In such cases I have but one remedy, viz. injecting into the holes spirits of wine in which corrosive sublimate has been dissolved. This is not so tedious a business as it may seem to be, as the spirit will often find its way from one hole to another, so that if half a dozen holes be judiciously selected, the poison will penetrate the whole piece of wood, kill all the insect inhabitants, and render it for ever impervious to their attack. The above-mentioned bow cost me but little trouble. I first shook out the greater part of the yellow powder, and then, placing the bow perpendicularly, injected the spirit into several holes at the upper end. The effect was magical. The little beetles came out of the holes in all directions, and not one survived the touch of the poisoned spirit; many of them, indeed, dying before they could force themselves completely out of the holes. The ticking of the deathwatch is, in fact, the call of the *anobium* to its mate, and as the insect is always found in old woods, it is very evident why the deathwatch is always heard in old houses. There is, by the way, a species of cockroach which acts in a similar manner, and generally disports itself on board ship, where the sailors know it by the name of 'Drummer.'"

The earliest account[*] we can find of the use of corrosive sublimate to destroy worms in woods is a few words mentioned, in 1705, by M. Homberg, French Academian. In

[*] 'Histoire de l'Académie,' p. 38. 1705. See also M. Maxime Paulet's communication to the Academy, 27th April, 1874.

that year he stated that a person of position in Provence, France, knew how to make a parquet floor which would resist the worm, viz. by soaking the wood in water in which corrosive sublimate had been mixed, and this process he had always found to be very successful.

Herr Temmnick preserved his books from the *anobium* by dipping them in a solution of quassia. Except on a small scale, however, the saturation of furniture seems scarcely practicable. Fumigation seems, however, to be more available. For small objects, the practice adopted at the Bodleian Library, Oxford, on Professor Westwood's recommendation, appears good, viz. to enclose a number of volumes in a box, shutting quite close, and placing a small quantity of benzine in a saucer at the bottom of the case. The same plan might be adopted with small ornamental wood-works, enclosing them in glass cases shut as nearly air-tight as possible.

The Report of the Commission appointed by the Department of Science and Art to inquire into the causes of decay in wood carvings, and the means of preventing and remedying the effects of such decay, which was published in 1864, states that the action of the worm in wood carvings may be arrested, and the worm itself destroyed, by vaporization, more especially by the vapour of benzine; and that, after the worm has been destroyed, further attacks from it can be prevented by treating the carved work with a solution of chloride of mercury, either in methylated spirits of wine, or parchment size, according to the surface character of the carving or wood-work; the strength of the solution in each case being 60 grains of the chloride of mercury to a pint of fluid, whether spirits of wine or parchment size. The carving or wood-work should be placed in a box, made as air-tight as possible, but with means of renewing the benzine placed in saucers from time to time as it evaporates without opening the lid of the box. Gilded carved work and panels on which pictures have been painted, and which have been attacked by the worm, can only be treated by applying the fumes of the benzine to the back of the pictures or gilded carved work: there is no reason to suppose that the vapour of the benzine would influence either the gilding of the one or the colours of the other.

The process should always be carried out during the spring and early summer months, according to the state of the temperature and the observations of those in charge of the carved or other work, as to the action of the worm, which is manifested by the fine dust falling from the worm-holes, crevices, &c.

Mr. Henry Crace was engaged in 1855 to restore some of the wood carvings in the Mercers' Hall, London, which had been perfectly honeycombed by a small brown beetle about the size of a pin's head. The carvings being first washed, a number of holes were bored in the back by a gimlet, and also into every projecting piece of fruit and leaves on the face. The whole was then placed in a long trough, 15 inches deep, and covered with a solution, prepared in the following manner:—16 gallons of linseed oil, with 2 lbs. of litharge finely ground, 1 lb. of camphor, and 2 lbs. of red lead, were boiled for six hours, being well stirred the whole time; 6 lbs. of bees'-wax was then dissolved in a gallon of spirits of turpentine, and the whole mixed while warm thoroughly together.

In this solution the carving remained for twenty-four hours. When taken out the face was kept downwards, that the oil in the holes might soak down to the face of the carving. The dust was allowed to remain to form a substance for the future support of the wood, and as it became saturated with the oil it increased in bulk, and rendered the carving perfectly solid.

No insect has since been found to touch these carvings, as they could not subsist on this composition.

In 1855 the carvings of Grinling Gibbons, at Belton House, were in such a condition as to render it absolutely necessary that something should be done to prevent their complete destruction. To this end they were placed in the hands of Mr. W. G. Rogers, who undertook to experiment upon their restoration. This gentleman reported that the first step he took was to have the various pieces photographed, as a means of recording the position of each detail of the ornamentation, &c. The whole of the works were in a serious state of decay, portions being completely honeycombed by the worm. In order to destroy or prevent any future development of the insect within the wood, Mr. Rogers caused the whole to be saturated with a strong solution of corrosive sublimate (bi-chloride of mercury) in water. The colour of the wood, however, suffered so seriously by the action of the mercury that it was found necessary to adopt some means of restoring the original tint. (It gives a dark colour to the wood, which is caused by the metal contained in the sublimate.) This was effected by ammonia in the first instance, and subsequently by a slight treatment with muriatic acid. After this the interior of the wood was injected with vegetable gum and gelatine, in order to fill up the worm-holes and strengthen the fabric of the carvings. A varnish of resin, dissolved in spirits of wine, was afterwards spread on the surface, and then the dismembered pieces were put together in conformity with the photographs taken, as records, prior to the work of restoration having been commenced.

In order to ascertain the present condition of these carvings, seven years after the operations detailed had been completed Mr. Rogers communicated with the Hon Edward Cust, one of the trustees of the Earl Brownlow, who desired him to communicate with the clerk of the works at Belton, Mr. G. A. Lowe. Mr. Lowe, in writing to Mr. Rogers, informed him that "there is never any appearance of worm dust from the very beautiful carving by Gibbons since you preserved it some years back."

Mr. Rogers stated, at a meeting of the Royal Institute of British Architects, a few years since, that similar carvings at Ditton Park, Cashiobury, and Trinity College, Oxford, are in a state of decay, the surface or skin, in some instances, being covered with a deceptive white vegetable bloom, which assists in completing the work of destruction.

Painting hastens the work of destruction. In the library of Trinity College, Cambridge, some of the finest carved work at some former time was thickly painted over, preventing the escape of the insects within, which were compelled to feed on the last bit of woody fibre, leaving nothing but the *skeleton* of what it once was. At Cashiobury, where can be seen room after room of the finest of Gibbons' work, all this charming carving (about thirty years ago) was covered over and loaded with a thick brown paint and heavy varnish, destroying all the delicate feathering of the birds and veining of the leafage, the repairs being done in plaster or a composition. Flowers, each about the size of an orange, were thus left with nothing but a skin of dust, with just enough fibre left to save them from collapsing in the handling. All the glorious work of Gibbons in the chapel of Trinity College, Oxford, was some years since covered with a dirty, undrying oil.

We dislike painters who paint carvings as much as the servant who applied to Mrs. H—— (wife of the celebrated landscape painter) for an "appointment" as cook, and having ascertained that the master of the house was a "painter," remarked, "I cannot take the situation, ma'am, as I have never lived in a *tradesman's family*."

It is a difficult process to remove paint from carvings, as it is not possible to scour and wash it off in the ordinary way: it must be eaten off by an alkaline solution.

With reference to the restoration of carvings which have not been painted, but only blackened by time, they must be scoured by the careful hand of an experienced man.

Mr. Penrose, the present architect to the Dean and Chapter of St. Paul's, a short time ago examined the beautiful carvings in St. Paul's Cathedral, and he was able to state that **they** have **not** hitherto **been attacked by** worms. " Some portions had **been** broken **by violence,** but the state of preservation was **marvellous." Mr. Rogers** also observed **that** " he was greatly **and** agreeably surprised—contrary to his expectations—to find the carvings in St. Paul's **in** so good **a state of** preservation, and so free from the attacks of insects; but such was undoubtedly the fact. How it was so he was not able **to say." Why** was this? Well, Sir Christopher Wren was **a wise man,** and when he erected St. Paul's Cathedral, he engaged an experienced mason to remain at the Portland stone quarries, **whose duty was** to select every block of **stone** for the Cathedral, and when **it arrived in** London **it was** placed *on its natural bed*. The good results of **this** precaution **can now be seen in the** good preservation of the stone at the present time. If he **was so** careful **of the** stone for the walls, no doubt equal care **was** taken **in the** selection of the **wood for** carvings. Besides, **the** *instructions** to the commissioners for rebuilding St. Paul's **were** drawn up with **a view of** preventing **decay.** The following **is** an extract from these instructions:

" And to call to your Aid and Assistance such **skilful** Artists, Officers, and Workmen as ye shall think fit, **and** to appoint **each of** them his several Charge and Employment; to minister **to such Artists and** Officers, **and to** all and every other **person** and persons to be imployed **in the** said service, to whom you shall **think meet,** such Oath or

---

* **Their Majesties'** Commission **for** the Rebuilding of **the Cathedral Church of St.** Paul, in London. London : Printed by Benjamin Motte. 1692.

Oaths for the due performance of their several Duties, Employments, Offices, Charges and Trusts to them or any of them to be committed as shall by you be thought reasonable and convenient; and out of such Money as shall be received for this Work, to allow to them, and every of them, such Salaries, Wages, and Rewards respectively as to you shall seem fitting and proportionable to their Employment and the Service they shall do." *

Sir Christopher Wren was descended from Dutch ancestors: he was building for a Dutch king, and we therefore perceive the reason why so much Dutch wainscot was introduced by him into England.

It seems a great pity that the beautiful carvings of Grinling Gibbons and others should be allowed to go to decay for want of proper attention. Why should this be? We are acquainted with some of Gibbons' carvings, particularly those in St. James's Church, Piccadilly, London; but whether they are in a state of decay unknown to any one, whether any one looks after them, or whether it is "nobody's business" to do so, we cannot say. Every now and then the owner of some beautiful wooden carvings suddenly becomes acquainted with the fact that they are thoroughly riddled through by worms, and instead of having them looked after, they are pointed at as curiosities. Even the makers of "old furniture" take care that it shall be bored all over, to imitate the borings of worms.

But what can be the cause of this decay? It must arise from one of two causes; or, it may arise from both, viz. either the wood was not seasoned when fixed; or else

* Workmen would now think this clause a striking one.

the quality and description of the wood for carving purposes was not attended to. There cannot be smoke without cause, and worms cannot exist unless a suitable habitation is first provided for them. Hard white oak is close grained, and will scarcely admit moisture; whilst on the other hand the soft foxy-coloured oak from some parts of Lincolnshire, and other places, is so porous as to imbibe it easily and retain it; and consequently is liable to early decay: in fine, the heart of this is scarcely equal to the sap of hard white oak.

The English woods least liable to the worm for carvings are cedar, walnut, plane, and cypress; those most liable are beech, pear, alder, ash, birch, sycamore, and lime. All the fine carvings at Blenheim, Kidlington, and Wimpole are in *yellow deal*, while in the age just before nothing but lime-tree and soft wainscot were used. The beautiful carvings of Gibbons, in the chapel of Trinity College, Oxford, are wrought in costly scented cedar and rich dark oak; those in Trinity College, Cambridge, in white lime-tree wood.

There is no doubt that wood to be used for carving should be hot, pungent, and bitter: thoroughly obnoxious to wood-destroying insects. If we cannot obtain this wood in England, we certainly can abroad, and one ship-load would last a long time for such purposes. Take, for instance, the *Jarrah* of Western Australia; the *Determa*, the *Cabacalli*, and *Kakatilly*, of British Guiana; and the *Sepe*, of Trinidad: these woods are much valued where they grow, and no insects ever attack them. We do not say that they are suitable for wood carvings, but they might be tried, and we certainly know they are not likely to be worm-eaten at the end of a few years. They need not be discarded on account of their hardness; boxwood is hard, but some good carvings have been executed with boxwood. We can relate an anecdote about this wood. On 3rd June, 1867, Mr. W. G. Rogers, the celebrated wood-carver (would that he were alive now to read these words), was asked, at the Royal Institute of British Architects, if boxwood is objectionable for wood carvings,* and he did not reply to the question; if he had given his opinion it would have been a valuable one, coming from such an authority. We must therefore get Mr. Rogers' opinion of this wood in another way. If the reader will refer to the "Reports by the Juries," English Exhibition, 1851, vol. ii., page 1555, he will find the following words:

"W. G. ROGERS, of London.—A cradle executed in boxwood for Her Majesty Queen Victoria, and richly ornamented with carved reliefs; also, a group of musical instruments, among which may be especially noticed a violin. These works show an extraordinary dexterity in the treatment of the material, and the ornaments of the cradle are in excellent taste. Prize medal."

We have already referred to the Report of the Commission on the Decay of Wood Carvings, and as this report is now rather difficult to be obtained, we propose condensing some extracts from it, which may prove of value to the reader.

* See lecture by Mr. W. G. Rogers, "On the Carvings of Grinling Gibbons," delivered at the Royal Institute of British Architects, 3rd June, 1867.

Of the three species of beetles injurious to furniture and carved work, the first, *Ptilinus pectinicorius* is about one-fourth of an inch in length, and the male is distinguished by its beautiful branched antennæ; the second, *Anobium striatum*, which is by far the commonest and most destructive, is about one-eighth of an inch long and of a brown colour, with rows of small dots down the back; and the third, *Anobium tessellatum*, is about one-third to one-fourth of an inch long, the back varied with lighter and darker shades of brown scales.

These insects are produced from eggs deposited by the females in crevices of the woodwork, from which are hatched small white fleshy grubs resembling the grubs of the cockchafer in miniature, which generally lie curled upon their sides, making very little use of their six small feet fixed near the head; it is in this state that the insect is chiefly injurious, although the perfect insect itself also feeds on the wood. These grubs make their burrows generally in the direction of the fibre of the wood; but when it becomes thoroughly dry and old, they burrow in all directions.

When full grown they cease eating, cast off their larva skins, and appear as inactive chrysalids with all the limbs lying upon the breast inclosed in little sheaths: after a short time the perfect insect bursts forth.

The appearance of the insects in the perfect state takes place uniformly during the first hot days at the beginning of summer. Where they take a liking to a piece of woodwork, they seem to devour every particle of it, and as the perfect insects possess large wings beneath their hard wing sheaths, they are often seen flying in the hot sunshine out of doors, evidently in search of suitable woodwork for themselves and their progeny.

Experiments were made by Mr. G. Wallis, Secretary to the Commission, with a view of ascertaining the best means of stopping the decay when commenced. The course pursued, as well as the results arrived at, will be best illustrated by a summary of Mr. Wallis's report on the subject.

The experiments may be placed under two heads, viz. Vaporization and Saturation.

I. Vaporization.

At the end of April, 1863, when, from the appearance of certain specimens of carved work, the worm appeared to be developed and active, a large glass case, made as air-tight as circumstances would permit, was filled with examples of furniture, &c.

The bottom of this case was covered with white paper, and the specimens of woodwork were raised above the surface by placing blocks of wood at convenient points. This insured the free circulation of the vapour over the whole surface of the objects. A dozen small saucers, with pieces of sponge saturated with carbolic acid, were distributed about the bottom of the case.

The raising of the objects on blocks of wood facilitated the placing of these saucers at any desirable point.

The carbolic acid was, in this experiment, renewed

every three or four days for a month, and a strong vapour pervaded the case for that period, during which there was no appearance of worms, dead or alive. At the end of May the saucers were removed, and the doors of the case thrown open, so that it might be well ventilated and cleared of vapour, after which it was closed again; but the saucers were not replaced. This closing of the case without using the vapour was to prevent the escape of any beetles which might make their appearance, in the event of the vapour of the creosote not having destroyed the worms. About the middle of June, a fortnight after the case was closed again, beetles were seen crawling upon the white paper with which the bottom was covered. These beetles would, no doubt, deposit their ova in the usual course, as they could not escape, and a considerable number of them were found dead upon the white paper with which the surface underneath the carved work was covered.

In order to test the efficacy of chloroform and benzine, two small glass cases, as nearly air-tight as possible, were selected, in which were arranged early in May specimens of ornamental woodwork, all more or less in bad condition from the worm. The bottom of each, as in the previous experiments, was covered with white paper; and the objects to be acted upon raised upon small blocks of wood. In one case chloroform was used, and in the other benzine in a similar manner to the carbolic acid, i. e. by placing small pieces of sponge in saucers and saturating them with the liquid, using five saucers in each case. Both the chloroform and the benzine had to be renewed much oftener than the carbolic acid, as the liquid evaporated much quicker.

Within a week after the experiment commenced it was evident that the action of the chloroform had destroyed the worms as they came to maturity, and in a fortnight all the specimens of carved work having been taken from the case, and the dust produced by the action of the worms shaken out, a number of dead ones were found, as also some dead beetles; but these were evidently those of past seasons remaining in the crevices of the woodwork.

On examining the specimens of carved work placed in the case treated with benzine, there was no appearance of worms or beetles dead or alive.

The two cases, with their contents, were then kept open for a week, and thoroughly ventilated to clear them as far as possible of all fumes of either chloroform or benzine.

After this they were closed again, being then free from all traces of vapour, and were not opened for some months. Throughout the summer, the temperature being the same as that under which beetles appeared in the case treated with carbolic acid, no traces of worms or insects were visible, nor could the remains of any be discovered on the white paper, with which the lower surface of each case was covered.

It would appear then, as far as vaporization is concerned, that the action of the vapour of carbolic acid is not sufficient; in fact, it is sluggish and heavy, whilst chloroform and benzine are volatile and penetrating. The experiment with chloroform appears to prove that the vapour kills the worm, and, as no beetles appeared in the case during the summer, it may be inferred that it *killed all* the worms within its influence.

From the pungency and penetrative action of the benzine, as also its volatile character and the fact that no life in the form of either worm or beetle was manifested in the case in which it was used, it seems fair to infer that it is more effective than even the chloroform.

Vaporization on a large scale might be adopted by having a room made as air-tight as possible, stopping up the chimney, pasting the window frames, &c., and placing infected furniture in the room, burning brimstone, or filling the room with fumes of prussic acid, chloroform, or benzine. It would have to be practised at the time when the perfect beetles made their appearance; their destruction at that time involving, of course, the prevention of further injury by their progeny.

II. Saturation.

The experiments made with bi-chloride of mercury (corrosive sublimate) and methylated spirits of wine were not so successful as by vaporization, on account of the woodwork when dry (after having been saturated with the solution) having a varnished appearance.

No experiment as to the effect of saturation in a solution of corrosive sublimate in water was made: 1st, because of the great risk to delicate carvings or pieces of furniture by their immersion in water, or the bringing up of the grain of the wood by treatment with a brush; and 2nd, because the vaporization by benzine appeared to be quite sufficient to destroy the larvæ.

Before terminating this chapter, we trust a few words about carvers and carvings will not be out of place.

There are two kinds of carvers, the *house carvers* and the *ship carvers;* the former are used to flat and square surfaces, the latter to the rake or fay, as was the old term.

About the period of Louis XIV. Malines was remarkable for its wood carvers, and the inhabitants might be seen sitting at their doors in the streets, plying their art in the same manner as now in many of the German and Swiss towns. Many works of art and decoration of Flemish origin are still preserved in England;[*] the works of Flemish carvers in wood were in great esteem, and there are numerous fine examples in the churches of Norfolk, and other parts of England which may be regarded as their productions. Evelyn remarked that Gibbons came from the Low Countries.

Grinling Gibbons created a school of carvers in England, and adopted a style and manner in building up his fruit and flowers to produce a grand effect. He chose but very few varieties of these out of his own garden, and it is wonderful how he varied and played with those few. He originated a peculiar description of light interlacing

---

[*] Paper by M. de Laperier, of the Belgian Legation, read at a meeting of the Society of Antiquaries, relative to Flemish origin of English carving.

scroll-work, which is to be met with in his best works; no one has successfully attempted to carry it on since his time. There are several examples at Belton, and in the chapel and state rooms at Chatsworth, in the fine trophies at Kirthington Park; but the upper part of the reredos of St. James's, Piccadilly, is a marvellous specimen."* The horizontal bands on the great organ in St. Paul's Cathedral are the perfection of this character of foliated scroll-work.†

Gibbons' carvings have a loose freedom about them. At Chatsworth he educated his workmen, who partook of his inspiration. There is a great deal of his work scattered over the rooms, great hall, and staircase of Lyme Hall, near Disley, which was erected under the direction of Sir C. Wren. It was executed by the persons who were employed at Chatsworth, and took nine years to complete.

At Blenheim there are some fine specimens of Chippendale's work, but what it all means is a mystery. Such a mixture of scraggy birds, and flowers cut into shreds, pagodas, and rustic waterfalls—all this fine workmanship employed to produce nothing but an incongruous whole of absurd objects. There is a leading line in all these works, indicating what the old carvers used to call the C and G style; because if you attempt to draw it, it will resolve itself into these two letters. There is also the S and G style.

* The large pulpit is not from the design of Sir Christopher Wren, nor is the carving by Grinling Gibbons.
† See engraving in the 'Art Journal,' 1866.

Abolish painting and we shall again have some fine house carvers.

We have already given the conclusions at which the Commission appointed by the Department of Science and Art arrived, as to the prevention of decay or attack by these insects, and will now conclude this chapter by quoting Dean Swift's recipe for getting rid of the Anobium or Death watch:—

> "But a kettle of scalding hot water injected,
> Infallibly cures the timber affected;
> The omen is broken, the danger is over,
> The maggot will die, the sick will recover."

CHAPTER X.

SUMMARY OF CURATIVE PROCESSES.

The following summary of the most approved formula for preventing and curing the evils of rot is prepared from the works of Tredgold and Wylson; some other more modern receipts have been added from 'The Builder,' 'Architect,' 'Building News,' and other professional periodical publications. Discretion in their use is recommended, and in serious cases we decidedly recommend consulting a professional man who is well acquainted with the subject.

TO PRESERVE WOODWORKS THAT ARE EXPOSED TO WET OR DAMP.

1. For those of an extensive nature, such as bridges, &c. The Hollanders use for the preservation of their sluices and floodgates, drawbridges, and other huge beams of timber exposed to the sun and constant changes of the atmosphere, a certain mixture of pitch and tar, upon which they strew small pieces of shell broken finely—almost to a powder—and mixed with sea-sand, and the scales of iron, small and sifted, which incrusts and preserves it effectually.

2. A paint composed of sub-sulphate of iron (the

refuse of the copperas pans), ground up with any common oil, and thinned with coal-tar oil, having a little pitch dissolved in it, is flexible, and impervious to moisture.

3. Linseed oil and tar, in equal parts, well boiled together, and used while boiling, rubbed plentifully over the work while hot, after being scorched all over by wood burnt under it, strikes half an inch or more into the wood, closes the pores, and makes it hard and durable either under or out of water.

4. For fences, and similar works, a coating of coal-tar, sanded over; or, boil together one gallon of coal-tar and $2\frac{1}{2}$ lb. of white copperas, and lay it on hot.

TO PREVENT ROT.

1. Thoroughly season the wood before fixing, and when fixed, have a proper ventilation all round it.

2. Charring, after seasoning, will fortify timber against infection, so will a coating of coal-tar.

TO CURE INCIPIENT DRY ROT.

1. If very much infected, remove the timber, and replace with new.

2. A pure solution of corrosive sublimate in water, in the proportion of an ounce to a gallon, used hot, is considered a very effectual wash.

3. A solution of sulphate of copper, half a pound to the gallon of water, laid on hot.

4. A strong solution of sulphate of iron; this is not so good as sulphate of copper.

5. A strong solution of sulphates of iron and copper in

equal parts, half a pound of the sulphates to one and a half gallon of water.

6. Paraffin oil, the commonest and cheapest naphtha and oil, or a little resinous matter dissolved and mixed with oil, will stay the wet rot.

7. Remove the parts affected, and wash with dilute sulphuric acid the remaining woodwork.

8. Dissolve one pound of sulphate of copper in one gallon of boiling water, then add $1\frac{1}{4}$ lb. of sulphuric acid in six gallons of water, and apply hot.

TO PREVENT WORMS IN TIMBER.

1. Anointing with an oil produced by the immersion of sulphur in aquafortis (nitric acid) distilled to dryness, and exposed to dissolve in the air.

2. Soaking in an infusion of quassia renders the wood bitter.

3. Creosoting timber, if the smell is not objectionable.

4. Anointing the timber with oil of spike, juniper, or turpentine, is efficacious in some degree.

5. For small articles, cover freely with copal varnish in linseed oil.

TO PREVENT WORMS IN MARINE BUILDING.

1. A mixture of lime, sulphur, and colocynth with pitch.

2. Saturating the pores with coal-tar, either alone or after a solution of corrosive sublimate has been soaked and dried into the wood.

3. Sheathing with thin copper over tarred felt is

esteemed the best protection for the bottoms of ships for all marine animals; the joints should be stopped with tarred oakum.

4. Studding the parts under water with short broad-headed nails.

TO DESTROY WORMS IN CARVINGS.

1. Fumigate the wood with benzine.

2. Saturate the wood with a strong solution of corrosive sublimate: if used for carvings, the colour should be restored by ammonia, and then by a weak solution of hydrochloric acid; the holes may be stopped up with gum and gelatine, and a varnish of resin dissolved in spirits of wine should afterwards be applied to the surface.

3. Whale-oil and poisonous ointments have been found of service.

The wood should be carefully brushed before being operated upon.

TO DESTROY ANTS AND INSECTS IN WOOD.

1. Corrosive sublimate is an effectual poison to them.

2. Oils, especially essential oils, are good preventives.

3. Cajeput-oil has been proved effectual for destroying the red ant.

4. Payne's, Bethell's, and Burnett's processes are said to be proof against the white ant of India.

5. Dust the parts with pounded quicklime, and then water them with the ammoniacal liquor of gas-works, when the ammonia will be instantly disengaged by the quick-lime, and this is destructive to insect life.

6. For the black ant, use powdered borax; or smear the parts frequented by them with petroleum oil; or syringe their nests with fluoric acid or spirits of tar, to be done with a leaden syringe; or pour down the holes boiling water to destroy their nests, and then stop up the holes with cement. Ants dislike arsenic, camphor, and creosote.

The preceding remedies are not by any means given with the intention of superseding the previous chapters, which should be carefully studied by those who wish to acquire a moderate knowledge of the subjects.

CHAPTER XI.

GENERAL REMARKS AND CONCLUSION.

Our task is nearly completed: we have but few general remarks to make.

The decay of wooden sleepers, posts, &c., on our railways and the destruction of timber piles by worms have been the causes of directing the attention of engineers to the preservation of timber. Most of our leading engineers now have the greater portion of the timber used in their works either creosoted or injected with chloride of zinc. Architects, as a rule, do not, unfortunately, adopt any process for preserving timber from rot and decay; and have practically no guarantee that timber used in their works has been thoroughly seasoned: posterity will not thank them for this, and yet they are not solely to blame. The fault in a great measure rests with the public, who require buildings to be erected at the least cost and in the shortest possible time. Moreover, the works executed by our leading builders are so extensive, that they have no room in their yards for large piles of timber to lie and season; and even if they had room it is doubtful if they would allow so much material, representing money, to remain idle. We are acquainted with one instance where a London architect, about a dozen years ago, erected a public building. The front of the reporters' gallery was formed of oak panelling; and within a year after the completion of the building narrow slips or tongues of wood had to be let in in several places to fill up the holes formed by the shrinkage of the panels. Similar cases to this are by no means rare. We can quote another instance of unseasoned wood. A range of workshops was erected a few years since in South London; the principals of the roof were not ceiled; almost before the building was finished the upper floor was occupied by a battalion of workwomen. The heat of the room (the ventilation being defective) soon had an effect upon the tie-beams, but one beam, which we imagine was unseasoned, in consequence of large shakes and splits, had to be taken out and replaced with new. We will (as a lawyer would say) cite one more case. A church in Surrey required some extensive repairs to the roof: an architect and a builder were employed, and the necessary works were done. Within four years dry rot has made its appearance on the new timbers of the roof (not an airtight one). One of the churchwardens, on consulting us last year (1874) as to the best means of stopping the rot, energetically remarked, "Who is responsible to us for this, the architect or the builder?" Charles Dickens, in his edition of 'Bleak House' in 1868, wrote, with reference to long Chancery suits, "If I wanted other authorities for Jarndyce and Jarndyce, I could rain them on these pages." We are able to make a similar remark with reference to any more instances of dry rot. According to the 7th chapter of the First Book of Kings, "Solomon was building his own house thirteen years:" we cannot spare so much time now-a-days over the erection of a house, but that is no reason why our timber should not be naturally or artificially seasoned.

If we cannot obtain naturally seasoned timber, by all means let us have artificially seasoned wood. Tredgold, in his Report on Langton's system,* nearly arrived at the secret. We will quote a few words from his Report:

"Mr. Langton having discovered a new method of seasoning timber .... by which the time necessary to season green timber, and render it fit for use, is only about twice as many weeks as the ordinary process requires years; .... it is more economical, and locks up less capital than the common method."

We believe we may say that the number of our public buildings which have been erected during the present century with artificially prepared timber can be counted on our eight fingers (without troubling our thumbs) and not exceed that number;† and yet we hear of dry rot in the great dome of the Bank of England and other buildings without profiting by the events. We should like to know if the wooden dome of St. Paul's Cathedral is safe from dry rot, (the domes at the Pantheon and the Halle-au-Blé at Paris were affected,) *and plumbers' fires.*

It is evident that a preservative process, thoroughly suitable for everyday use and applicable to buildings,

---

* See Tredgold's Report on this process, May 2, 1828.

† See Bartholomew's 'Specifications,' and Professor Donaldson's valuable work on 'Specifications,' which comprises many examples by modern architects. The usual clause is: "The timber to be well seasoned (is it?), free from large knots, shakes, and other defects."

has yet to be invented: it should be cheap, should render wood uninflammable, should preserve the wood from decay and dry rot, *should not harden the wood until some time after its application*, and should be colourless and invisible. The invention of such a process will require careful thought and experiments, for it appears to us that the whole theory of any successful plan for the prevention of the dry rot must resolve itself into the *solidifying or coagulation of albumen*: this means hardening the sap-wood, and causing increased difficulty in working the wood. We can easily illustrate our remarks, by quoting one of the latest patents for preserving timber, which has recently been made public. It is the invention of a gentleman living in England, who has discovered a means of making wood uninflammable, preventing dry rot and decay, and rendering white and yellow pine, both in hardness and appearance, like teak and oak. We have no objection to its rendering wood uninflammable, providing it does not "hurt" the wood; but can the reader believe that any architect, in erecting a moderate-sized villa, would specify that all the joiners' work, staircases, window-frames and sashes, architraves, skirtings, doors, &c., must be formed of wood *as hard as teak;* or rather, can the reader imagine the architect's client would be agreeable to pay the greatly increased cost for the extra labour involved. We do not think this invention will ever be used, at least to any extent, in buildings.

Much yet remains to be done with regard to uninflammable wood for buildings: we think the matter should be dealt with (with reference to joists, floor boards, partitions, doors, staircases, roof timbers, &c.) by a new Buildings Act of Parliament. Stone and iron will not burn, but they are not fire-resisting: brick, artificial stone, and incombustible wood will give us all we desire; the details may be difficult of arrangement, but builders would comply with them if they were imperatively required. At present our houses are formed of brick walls, every room being separated vertically and horizontally from the adjoining rooms by combustible wooden walls. A street built up of fire-proof buildings would be a novelty. The whole subject requires to be dealt with thoroughly, for while we have combustible wooden floors, partitions, &c., we cannot at the same time have a fire-proof building. We have not been able to spare the space, or else we should have devoted a long chapter to this subject; a superficial consideration (such as alum and water) would have been practically useless.

In conclusion, we can only summarize our remarks on the cause of dry rot, by saying, "Season and ventilate," in every case: as to the cure, that is not so easy to deal with. If the reader has ever had a decayed tooth aching, a friend has probably said, "Have it out;" and we say, wherever there is a piece of timber decayed in a building which can be removed, "Have it out, and stop up with new;" and in so advising we are merely following the advice to be found in a good old volume, which has never yet been equalled, and which says:

"And, behold, if the plague be in the walls of the house with hollow strakes, greenish or reddish, which in sight are lower than the wall; . . . . Then the priest shall command that they take away the stones in which the plague is, and they shall cast them into an unclean place without the city: And he shall cause the house to be scraped within round about, and they shall pour out the dust that they scrape off without the city into an unclean place: And they shall take other stones, and put them in the place of those stones; and he shall take other mortar, and shall plaister the house."—*Leviticus* **xiv. 37, 40, 41, and 42.**

This course will not, however, suit every case, for when the rot has spread in many directions, the best and cheapest course is to consult some professional man, well versed in the peculiarities of dry rot, before determining upon any remedy, for we have shown in the course of this work that the disease may arise from various causes; and it is not a difficult matter to select the wrong remedy, and thus increase the disease.

We trust the reader has found in this volume at least some hints which may be of service to him. A *new* house affected with dry rot is an unhealthy one to live in, and an *old* one is worse than the new; we mean the kind of house referred to in one line by an American poet, as follows:

"O'er whose unsteady floor, that sways and bends."
LONGFELLOW.

[INDEX.

INDEX.

	PAGE
ABEL's silicate of soda process	160
Academy of Sciences, Holland, report on sea-worms	235
Acetate of lead	226
,, iron and wood tar	130
Acid, carbolic	257, 276
,, fluoric	287
,, hydrochloric	286
,, hydro-fluo-silicic, and other substances	166
,, nitric	98, 285
,, pyroligneous	111, 144, 263
,, sulphuric	161, 285
,, vegetable	111
Age of trees, how to ascertain	9
Air, admission of, to prevent or cure rot	27, 171, 187, 284, 292
Alberti (L. B.), on seasoning wood	66, 75
Alcohol, in corrosive sublimate	263, 265, 266, 279
Alderson's (Captain), experiments with woods	127
Alkali, caustic	122
Alum, to prevent combustion	118
,, experiments with	119
,, and other substances	156, 166, 167
American method of preserving ships' masts	111
,, oak, inferior to English	40
Ammonia, to cure rot	118, 137
,, and other substances	131, 286
Amsterdam, built on piles	23
Annual rings in wood	8
Ants, black, how to destroy	287
,, white, description of	240
,, how to destroy	251, 286
,, in Australia, Bahia, and Pernambuco	245
,, in Batavia	247
,, in Brazil	244
,, in Ceylon and the Philippine Islands	246
,, in France and Japan	248

INDEX.

	PAGE
Ants, white, in India	251
,, in Jamaica	241
,, in Spain, Senegal, and Surinam	248
,, woods which resist	249
Armstrong's (J.), account of rotten floor	43
Arsenic	224, 252, 287
,, experiments with	167
,, and other substances	253
Asphalte, to keep out damp	179
Australian method of seasoning Jarrah wood	115
Baker's (J.), case of dry rot in Baltic wood	177
Ballast for railway sleepers	48, 138
Bank of England, dry rot in dome	42
Banks (Sir J.), on growth of fungi	44
Barium sulphide, to preserve wood	156
Barlow's patent process	102
,, on seasoning wood	78
Barnacles on timber piles	223, 226
Barry (Sir C.), on steaming wood	90
Baryta, and other substances	166
Basement stories with damp	23, 181, 182, 187
Bayonne, girder in church at	174
Beams, advantage of sawing	32
Bees, carpenter, destroy wood	240, 259
,, wax, and other substances	156
Beetles, in wood	262, 275
,, how to destroy	286
Belgian engineers prefer charred sleepers	96
Belidor, on felling trees	54
Belton House (Earl Brownlow's), beetles in carvings at	268, 281
Bentham (Sir S.), on drying oak	91
Benzine, to destroy wood beetles	266, 277, 286
Berkeley, on fungi	21
Bethell's (J.), patent creosoting process	130, 155, 224, 234, 286
,, ,, drying stoves	86
Binmer, on steaming and charring	99
Biot, on pressure process	144
Blenheim, state of carvings at	281
,, carvings in yellow deal at	273
Blood, and other substances	167
Bond timber, decay of in walls	45, 174
Borax, a receipt for black ants	287
,, and other substances	156

INDEX. 297

	PAGE
Boucherie's (Dr.), sulphate of copper process	146
Bourne's (J.), experiments with woods	254
Bowring's (Sir J.), account of ants in Obando	247
Boyden's (A.), remedies for dry rot	95, 112, 122
Brande (Dr.), on preserving woods	139, 142, 155
Bréant's patents	145
Brick dust, tar, &c., to preserve piles	228
Brimstone, beeswax, &c., to preserve wood	156
Brochard and Watteau's process	80
Browne's (Sir S.), experiments with piles	229
Brunel (Sir M. I.)	138, 139, 215, 228
Buffon	144, 198
Builders, bad	182, 202
Building, hints on	180
Burnett's (Sir F.), patent zinc process	140, 224, 254, 255, 286
Burt's experience of creosoted sleepers	137

CADET DE GASSICOURT's process for dry rot	144
Calomel, composition of	264
Calvert's caoutchouc process	162
Camphor disliked by ants	287
Canadian white spruce deals liable to warp	65
,, yellow wood liable to rot in damp situations	36, 43
Caoutchouc, solution of	162
,, and other substances	163
Carbolic acid, for wood beetles	257, 276
Carbonate of soda (Payne's process)	154
Carbonization by gas	97, 164
Carpenter bees destroy wood	240, 259
Carpenter (Dr.), on growth of fungi	43
Carvers, wood	280
Carvings destroyed by worms	266
,, how to clean	270
,, to destroy worms in	286
Cashiobury, carvings at, destroyed by beetles	269
Cement, to protect piles	227, 228
Ceylon, ants in	246
Chalk, and other substances	161
Champy's tallow process	144
Chapman (W.), on dry rot	25, 73, 112, 119, 122, 165, 167
Charcoal—*see* Oils, Whale, and Fish—to preserve wood	121
,, and other substances	157
Charpentier's hot air patent	80
Charring wood	95

INDEX.

	PAGE
Charring wood, when useful	160
,, and pitching	96
Chassloup Lambat's suggestion to prevent rot	163
Château of the Roques d'Oudres, girders at	174
Chatsworth, Gibbons' carvings at	281
Chelum terebrans destroy piles	219
Chemists prefer thin creosote	131
Chinese method of preserving wood	167
Chippendale's carvings	281
Chloride of calcium	146
,, of manganese	154
,, of sodium	164
,, of zinc—*see* Burnett's Process	
Chlorine gas, and other substances	123
Chloroform, for wood beetles	277
Chunam, and cocoa-nut oil	107
Church at Bayonne, fir girders in	174
,, of Holy Trinity, Cork, rot in vaults	39
,, in London, rot in roof	184
,, in Surrey, ,,	289
,, of St. Mark, Venice, rot in curb	176
,, of Old St. Pancras, London, rot in vaults	40
Cleghorn (Dr.), on creosoted sleepers	47, 136, 142
Coal Exchange, flooring of	81
,, tar	170, 233, 246, 256, 262
,, ,, and other substances	123, 284, 285
,, vessels last long	117
Cobley's patent lime process	166
Colocynth and quassia	263
,, and other substances	285
Colouring woods	108
Commission, report of, on carvings	266, 274
Cooke's (M. C.) instance of fungi	43
Copal varnish	191, 197
,, in linseed oil	285
Copper, red oxide of	161
,, prussiate of	146
,, sulphate of—*see* Sulphate of Copper	
,, nitrate of	226
,, sheathing against sea-worms	228
,, ,, and tarred felt	285
Copperas, and coal tar	284
,, to preserve ships	112, 226
Cork, for ends of brestsummers	174
Corrosive sublimate	123, 226, 264, 265, 285, 286

INDEX.

	PAGE
Corrosive sublimate and other substances	130, 155, 263, 265, 266, 279, 285
Covent Garden Theatre, dry rot in bond	175
Cow-dung mortar, and oils	251
Creosote (Bethell's patent),	118, 130, 133, 142, 165, 230, 236, 255, 257, 285, 287
,, vapour	145
,, and chloride of zinc	133
Crépin (M.), on creosoted wood	139, 236
Cryptogamia, or fungi	15
Cullen's process for dry rot	157
Dammer oil, and other substances	255
Damp	176, 177, 178, 181
,, a cause of decay in wood	22
,, rooms, how to ascertain	24
Darwin's process for dry rot	156
Daviller (A. C.), on felling trees	54
Davison and Symington's process	81
Davy (Sir H.), on corrosive sublimate	127, 263
Deals require long seasoning	64
,, how sometimes imported	35
Deane's (Sir T.), account of dry rot case	39
Decay of trees, symptoms of	33
De Lapparent's processes	73, 97, 163
Desiccating processes	81
Dickson (Dr.), on Kyan's process	130
,, (J.), on seasoning wood	75
Ditton Park, carvings destroyed at	269
Donaldson's (Prof. T. L.) account of dry rot case	42
Dondeine's paint	165
Dorsett and Blythe's copper process	151
Doswell's report on timber piles	232
Dram battens liable to rot	38
Dry rot, wet rot, and rot.	
,, appearances of	31, 35
,, causes of	24
,, danger of	34
,, how different from wet rot	14
,, proceeds according to temperature	29, 187
,, caused by bad building	182
,, ,, mortar	44, 173, 177
,, ,, damp brickwork	44, 182
,, ,, ,, ground	20, 21
,, ,, ,, stone	44
,, ,, heat and moisture	23

INDEX.

	PAGE
Dry rot caused by insufficient areas	178
" " " tarpaulins	184
" " joining different woods	176
" " kamptulicon	187
" " Keene's cement	188
" " oiled cloth	185
" " old trees	183
" " partial leaks	23
" " want of air 171, 172, 186, 187,	188
" " " proper drains and spouts	41
" increased by stoves	172
" *in ground*, under house at Hampstead	20
" *under foundations*, Norfolk House	176
" " " Grosvenor Place	176
" " *floor*, Stanmore Cottage	183
" " *hearthstone*	43
" " *pavement* at Basingstoke	43
" on *paved floor*, Westminster Hall	44
" *in vaults*, Old St. Pancras Church	40
" *on vaults*, Holy Trinity Church, Cork	39
" in cask in cellar	43
" " *basement floor* of house, Greenwich Frontispiece	
" " *ground floor* of houses 43, 177, 185, 186,	187
" " *first floor* of house, No. 29, Mincing Lane	187
" " *second floor* of house, No. 79, Gracechurch Street	187
" " *barn floor*	42
" on *floor* of house, No. 106, Fenchurch Street, London	186
" *in wood bond*, Covent Garden theatre	175
" " *damp closet*, or pantry	16
" " *wood lining* to walls—basement	
" " floor of house in the Temple, London	124
" " *brestsummer* of shop	42
" " *girder* of house (Earl of Mansfield's)	32
" " " building at Malta	32
" " *partition*, No. 16, Mark Lane, London	188
" " *roof*, church in London	184
" " " " Surrey	289
" " *curb of dome*, St. Mark's, Venice	176
" " *dome*, Bank of England	42
" " " Halle-au-Blé, Paris	42
" " " Panthéon, Paris	42
" " Society of Arts building, Adelphi	42
" " *field gates*	183
" " *foreign timber*	35
" " *paling*	125

	PAGE
Dry rot *in ships* 23, 26, 73, 93, 112, 114, 172	
„ prevented by seasoning	63
„ good, cheap, and easy remedy required	291
Du Hamel.. .. .. 66, 72, 144	
Duke of Devonshire's house, dry rot at	40
D'Uslaw's, Meyer, steam process	102
Dutch method of coating piles..	221

EARL Brownlow's house, beetles in carvings at	268
„ of Mansfield's house, rotten yellow fir girder at	32
Emerson's boiled oil process for rot	110
Endogenous stems, grow from within	4
Engineers, English 139, 288	
„ foreign, rules for sulphate of copper	151
„ „ „ creosote 131, 133	
Evelyn (Sir J.), on seasoning wood 53, 73, 75	
Exogenous stems, grow from without	4

FARADAY (Prof.), on corrosive sublimate 129, 263	
Felt, tarred, and copper sheathing	285
Fences, how to prevent them rotting 46, 161	
Fenchurch Street, No. 106, dry rot on floor	186
Feuchtwanger's (Dr.), water-glass for piles	226
Field gates, dry rot in	183
Fire-proof houses, cost of	143
„ „ necessity of	291
Flemish carvings in England	280
Flockton's wood tar process to preserve wood	130
Floor-cloths, injurious effects of	185
Floors, how to protect from worms	266
„ dry rot in .. 20, 39, 40, 42, 43, 44, 125, 176, 182, 183, 186, 187	
„ „ .. .. Frontispiece	
Fluoric acid, for the black ant	287
Fontenay's metallic soap, to preserve wood	165
Forestier's experiments with creosoted piles 139, 236	
Foundations, how to build	179
Fraser's (Capt. A.) paint for white ants	253
Fungi differ according to situation	22
„ explanation of the term	15
„ forms and strength of 31, 43	
„ production of 15, 18, 19, 20	
„ rapid growth of	44

	PAGE
GAMBIR composition for white ants	255
Garlic and vinegar for worms	106
Gas, carbonization of wood by	97, 164
„ chlorine, and other substances	123
Gibbons' (Grinling), carvings	260, 280
Glue, solution of, to preserve ships	112
„ and other substances	112, 122, 130
Gracechurch Street, No. 79, dry rot in second floor	187
Graham (Prof.), on Burnett's process	140
Grease, how to take it out of floor	191
Greenwich, rot in floor of house at	Frontispiece
Greville's (Dr.) description of fungi	21
Groo-groo worms in Surinam	247
Grosvenor Place, rotten planking in houses	176
Guibert's smoke process	93
HALES' (Dr.) oil and creosoting processes	111, 118
Halle-au-Blé, Paris, dry rot in dome of	42
Haller's (Dr.) analysis of a fungus	31
Hampstead, dry rot in ground of house at	20
Hancock's caoutchouc and oil process	162
Hartley's experiments with fire-proof house	120
Hawkshaw's opinion of Payne's process	155
Higgins' (Dr.) ammonia remedy for rot	118
House, fire-proof	120
„ „ cost of	143
„ badly erected	182, 202
Howe's experiments with posts	45
Humboldt, Baron, on damp rooms	24
INDESTRUCTIBLE Paint Company	195
Indian Woods	47, 134, 223, 250
Ingredients for preserving wood	168
Iron, cast, effect of sea-water on	230
„ muriate of	157
„ prussiate of	146
„ pyrolignite of	130, 146, 151, 156, 234
„ sulphate of	154, 157, 284
JACKSON'S preserving processes	111
„ (G.) experiments with white ants	254
Jagherry, or coarse Indian sugar, for mortar	253

INDEX.

	PAGE
Japanese method of treating graining	194
Jarrah wood, how seasoned	115
Johnson's (B.) account of rot in floor	42
Jones' (Major, R. E.) report on rotten beams	32
KAMPTULICON causes dry rot in floors	187
Kenwood, rotten fir girder at	32
Kidlington, carvings in yellow deal at	273
Kirthington Park, Gibbons' carvings at	281
Knabb's sulphate of copper process	152
Kœnig's opinion of sulphate of copper	152
Kyan's corrosive sublimate patent	123, 205, 223, 233
LAMPBLACK, and fish oil	108
Langton's extraction of sap process	101
Lead	173, 179, 200
„ and tarred rope for piles	228
„ oxide of, and other substances	123
Légé and Fleury-Pironnet's copper patent	149
Le Gras' manganese, zinc, and creosote patent	164
Lepisma worm destroys boats	221
Letellier's preserving processes	130, 165
Lewis' lime process	112, 116
Liebig (Baron) on decay of wood	19
Lime, to preserve wood	112, 116, 253, 286
„ and other substances	107, 117, 156, 157, 166, 255, 285
„ recarbonated, injurious to wood	116
„ water, to preserve ships	116, 122
„ „ „ basement joists	116
„ „ and sulphuric acid	156
„ vessels last long	116
Limnoria terebrans, description of	217
„ „ how it destroys piles	218
Linseed oil—see Oils	
Litharge „ „	
Logs, state of, on arrival in England	37
Lowestoft Harbour, creosoted piles in	230
Lukins' stove process	121
Lycoris fucata, destroys the Teredo navalis	237
Lyme Hall, carvings at	281
MACONOCHIE's suggestions for preserving wood	121, 145, 163
McMaster (B.), on decay of railway sleepers	47

McWilliam, on fungi	20, 22, 29
Makinson, on creosoted piles	231
Malta, rotten girders in building at	32
Manganese, and other substances	163, 165
Mann's (Capt.) and McPherson's (Capt.) experiments	255
Margary's patent sulphate of copper process	130, 150, 254
Mark Lane, No. 16, dry rot in partition at	188
Marshall (G.), on seasoning oak	69
Maun (G. O.), on sleepers, Pernambuco railway	138
Mecquenem's desiccating process	80
Mellis (J. C.), on creosoted wood	256
Melseun's experiments with ammonia	137
Mercer's Hall, decay of carvings at	267
Mercury, deuto-chloride of	165
,, bi-chloride—*see* Corrosive Sublimate	
Merulius lachrymans, dry rot fungus	21
Methods for seasoning wood	168
Methylated spirits of wine for carvings	279
Michigan Central Railroad bridge, dry rotten	185
Migneron's process	144
Miller's hot air process	102
Mincing Lane, No. 29, dry rot in first floor at	187
Moll's vapour of creosote process	145
Moon, age of, a guide for cutting trees	56
Mortar made with sea sand objectionable	113, 181
,, cow-dung and castor oil	251
Mud and other substances to preserve wood	253
Müenzing's manganese process	154
Mundic, to preserve wood	118
Muriate of iron (Toplis' process)	157
Nails, scupper, for piles	228, 286
Neumann, on seasoning wood	79, 117
Nichols (T.), on sand bath	116
Nitrate of copper for piles	226
Nitric acid, for worms	285
Norfolk House, rotten planking at	176
Norway white lowland deals warp	65
Nystrom's process, to prevent combustion	166
Oak, American, liable to rot	40
,, different qualities of	71
,, good and bad	25

INDEX.

Oak seasoning	69, 70, 90, 91
,, panelling, if not seasoned, shrinks	288
,, how to prevent splitting	106
Ohio fireproof paint	185
Oil, Arracan, to protect wood from ants	252
,, boiled, to preserve planks of ships	111
,, castor, with cow-dung mortar	251
,, cajeput, to protect wood from ants	247, 286
,, of cedar, to protect wood from worms	106
,, cocoa-nut, to preserve wood	107
,, ,, and other substances	107
,, dammer, and other substances	255
,, fish	108
,, ,, experiments with	108
,, ,, and other substances	108
,, linseed	106
,, ,, and other substances	106, 165, 268, 284, 285
,, olive	106
,, of juniper, to prevent worms	285
,, of mustard, to preserve wood	107
,, of spikenard	106, 285
,, of tar, and other substances	123, 155, 162
,, of tar—*see* Coal Tar	
,, palm, to preserve wood	106, 107
,, ,, and other substances	123
,, paraffin, to cure dry rot	285
,, petroleum, to preserve wood	109, 157, 169, 262, 287
,, ,, and sand	109
,, vegetable, best to preserve wood	106
,, whale	286
,, ,, renders wood brittle	106
,, ,, and other substances	106, 107
,, and other substances	156, 167
Oils, animal, render wood brittle	107
Oxford's patent	123
PAINTING, house, described	199
,, ,, causes rot	183, 185, 269
,, how to remove from carvings	270
Paling, rot in	185
Pallas' iron and lime process	117
Panthéon, Paris, dry rot in dome	42
Parkes' caoutchouc process	162
Parry's (Dr.) suggestion to prevent rot	156

	PAGE
Passez's caoutchouc in sulphur process	162
Pasteur, researches of	17
Patents, most successful patents	169
Payne's patent process	144, 154, 156, 223, 254
Peat moss, for seasoning wood	116
Penrose's report on carvings, St. Paul's Cathedral	271
Pepys, Memoirs of, account of rot in ships in	24
Pering on dry rot	25
Petersburgh deals, white and yellow	38, 66
Petroleum oil to prevent rot	109, 157, 169, 262, 287
Phillips (R.), on seasoning oak	70
Piles, timber	23, 96, 219, 221, 223, 226, 228, 285
,, ,, cased in iron	229
Pine, yellow, liable to rot	43
Pitch	96, 174, 224
,, and other substances	107, 159
Pith of tree, formation of	4
Pliny, on salt-water seasoning	72
Polyporus hybridus fungi	21
Porcher (Dr.), on seasoning wood	75
Posts, experiments with	45
,, in Norway, how preserved	173
,, burning ends to preserve	96, 98
,, where they decay	24
,, coating, to preserve	161
Potash, and other substances	166, 167
Price and Manby's drying stove	88
Pringle (Sir J.), on the strength of alum	119
Pritchard's report on sea-worms	156, 233
Processes, rules for successful	110
,, pressure and vacuum	168
Prussiate of copper (Boucherie's process)	146
,, of iron ,, ,,	146
Pyroligneous acid	111, 144, 263
Pyrolignite of iron	130, 146, 151, 234
,, ,, and oil of tar	156
Quassia	266, 285
,, and colocynth	263
Quatrefages' experiments	225, 242
Quicklime, if dry, preserves wood	116

INDEX. 307

RAILWAY sleepers, 47, 49, 74, 101, 103, 125, 134, 136, 138, 140, 143, 149, 151, 152, 251, 254

	PAGE
Rance's experiments with chloride of sodium	164
Randall (J.), on oxidating wood	98
Ransome's silicate of soda process	156, 227
Rats, how to get rid of	173
Reid's vegetable acid process	111
Remedies for white ants	286
,, for black ants	287
,, for dry rot	284
,, worms in carvings	286
,, ,, in piles	285
Renwick's vapour of creosote process	146
Resin, and other substances	122, 159, 161, 285
Robins, oleaginous vapour process	157
Rogers (W. J.), the wood carver	72, 268, 274
Rot, internal causes of	32
,, in timber, how to ascertain	33, 185
,, ,, to prevent	283
,, ,, to cure	284
SALT, bay, to preserve ships	114
,, common, to preserve ships	112
,, ,, to preserve railway sleepers	74
,, water, lime, &c., to preserve wood	73, 111
,, vessels last long	114
Saltpetre, to preserve ships	114
Salts, deliquescent, corrode metals	112
Sand and coal tar	284
,, and petroleum	109
,, bath	116
,, sea	113, 181
Sapwood in different woods	3
Saturating woods to resist beetles	279
Scott's (Col.) paint for ants	258
Sea salt and copperas	166
,, sand	113, 181
,, water, effect of, on iron	230
,, weed	113
,, worms	203
Seasoning by air, and exposure in stacks	64
,, ,, heated	80
,, by extraction of sap	101
,, ,, water, fresh	71

x 2

	PAGE
Seasoning by water, salt	73, 113
" " " " sea-weed, and sea-sand	115
" " " lime	73, 111
" " smoke	91
" " steaming and boiling	77
" " charring	99
" " gas	97, 164
" " sand bath	116
" " scorching and charring	95, 97
" " baking	79, 81, 86, 88, 94
" oak	69, 70, 72, 289
" second	103
Sea-worms, woods which resist	223
Selenite, experiments with	119
Shakes in wood	10, 249, 250
Shaw (Capt. E. M.), on admission of air	120, 171
Shield's remedy for white ants	245, 256
Ships	99, 111, 112, 114, 116, 117, 194, 251
" dry rot in	23, 26, 73, 93, 112, 114
Silicate of potash	155
" of soda	156, 160, 227
" " and lime	160
Silloway (T. W.), on seasoning wood	75, 92
Silver grain	6
Size for wood, why required	197
" and corrosive sublimate	266
Slating wall to keep out damp	177
Sleepers, see Railway Sleepers	
Smirke (Sir R.), on dry rot	20, 123
Smith's solution for wood beetles	264
Soap, experiments with	122
" metallic, to preserve wood	165
" yellow " "	165
" and other substances	253
Society of Arts building, dry rot in	42
Soda, carbonate of	155
Soluble glass	155
Southend pier, attacked by sea-worms	209
Spores, description of	15
Stains for woods	189, 197
Stanmore Cottage, dry rot in floor at	183
Steam	145, 168
" —see Seasoning by Steam	
Stephenson (Sir M.), on creosoted wood	134
Stevenson (R.), on timber piles	205, 217

INDEX. 309

	PAGE
St. James's Church, Piccadilly, carvings at	272, 281
St. Helena, experiments with woods at	256
St. Mark's, Venice, rotten curb of dome at	176
St. Paul's Cathedral, London	42, 271, 290
St. Preuve's steam process	80
Stove drying	79, 81, 86, 88, 94
Strength of timber	11
Strontia, and other substances	166
Sublimate—*see* Corrosive Sublimate	
Sulphate of **copper**	122, 146, 149, 150, 151, 161, 226, 284
„ „ and sulphuric acid	285
„ of **iron**	154, 157, 284
„ „ and other substances	117, 166, 284
Sulphur	163
„ in other substances	163, 285
Sulphuric acid	161, 285
Surinam, groo-groo worms in	247
Swift's, Dean, recipe for beetles	282
TALLOW bath for wood	144
Tar, and other substances	106, 130, 159, 228, 251, 284
Tarred rope, and lead for piles	228
Teak oil, to preserve **wood from ants**	259
„ chips, distilled	163
Temple of Diana, at Ephesus, built on charred piles	98
„ buildings, London, dry rot in	124
Tennant's (Sir E.) account of ants in **Ceylon**	246
„ „ bees „	260
Teredo navalis, description of	212
„ —*see* Worms, Sea	
Termites—*see* Ants, White	
Tie-beam, **instance of unseasoned**	289
Timber depreciates by **keeping too long**	64
Tissier's **hot air process**	102
Toplis' **sulphate of iron** process	157
Tredgold (T.), on seasoning wood	78, 101, 290
Treenails	26, 110, 118
Trees, symptoms of decay in	52
„ how **to** prepare for felling	61
„ when to fell	53, 54, 55, 58
Trinity College, Cambridge, carvings **at**	269, 273
„ Oxford „	269, 273
Truman's brewery, **seasoning** casks at	84
Turpentine **prevents rot**	36, 257, 263, 285
„ in corrosive sublimate	115

INDEX.

	PAGE
Uninflammable wood, good process required for	170, 291
Unseasoned oak panelling	288
,, roof principal	289
Vaporizing woods	276
Vapour of creosote process	145
Venice, built on piles	23
Vernet's fireproof method	167
Vessels in coal trade last long	117
,, in lime ,, ...	116
,, in salt ,, ...	114
Vinegar—*see* Garlic	
Vitriol, blue—*see* Sulphate of Copper	
,, green—*see* Sulphate of Iron	
Vitruvius on seasoning wood	75
Vulliamy (G.), on charring posts	96
Wade's suggestions for preserving wood	119, 122
Wainscot, Crown Riga	90
,, dry rot in	35, 125
,, how to cut oak for	70
,, unseasoned oak for	289
Wallis' experiments with beetles	276
Walnut juice for worms	263
Warburton's (H.) opinion of American oak	40
Warping of boards	66, 67
Water in wood	39, 67, 180
,, in church	29
,, glass to preserve piles	226
Watson's (Dr.) experiments with wood	67
Westwood's (Prof.) report on wood beetles	262
Wet rot, how caused	14, 28
Wimpole, carvings at	273
Wood bond decays	'175, 176
,, progress of decay in	19
,, (Rev. J.), on worms and ants	211, 265
Woods best when not painted	189
,, experiments with	46, 58, 67
,, french polished	192
,, white, improved by water seasoning	72
,, which resist beetles	273
,, ,, sea-worms	223

		PAGE
Woods which resist white ants		249
Woodcutters		55
,, tricks of Indian		11
,, tricks of, in Ceylon		114
Woody fibre, formation of..		2, 7
Worms, sea		203
,, how to prevent in wood		285
Wren (Sir C.)		23, 98, 221, 271

		PAGE
Zinc, chloride of—*see* Burnett's Process		
,, sulphate of..		122
,, white oxide of		226
,, and other substances		165

1885.

BOOKS RELATING
TO
APPLIED SCIENCE

PUBLISHED BY

E. & F. N. SPON,
LONDON: 125, STRAND.
NEW YORK: 35, MURRAY STREET.

A Pocket-Book for Chemists, Chemical Manufacturers, Metallurgists, Dyers, Distillers, Brewers, Sugar Refiners, Photographers, Students, etc., etc. By THOMAS BAYLEY, Assoc. R.C. Sc. Ireland, Analytical and Consulting Chemist and Assayer. Third edition, with additions, 437 pp., royal 32mo, roan, gilt edges, 5s.

SYNOPSIS OF CONTENTS:

Atomic Weights and Factors—Useful Data—Chemical Calculations—Rules for Indirect Analysis—Weights and Measures—Thermometers and Barometers—Chemical Physics—Boiling Points, etc.—Solubility of Substances—Methods of Obtaining Specific Gravity—Conversion of Hydrometers—Strength of Solutions by Specific Gravity—Analysis—Gas Analysis—Water Analysis—Qualitative Analysis and Reactions—Volumetric Analysis—Manipulation—Mineralogy — Assaying — Alcohol — Beer — Sugar — Miscellaneous Technological matter relating to Potash, Soda, Sulphuric Acid, Chlorine, Tar Products, Petroleum, Milk, Tallow, Photography, Prices, Wages, Appendix, etc., etc.

The Mechanician: A Treatise on the Construction and Manipulation of Tools, for the use and instruction of Young Engineers and Scientific Amateurs, comprising the Arts of Blacksmithing and Forging; the Construction and Manufacture of Hand Tools, and the various Methods of Using and Grinding them; the Construction of Machine Tools, and how to work them; Machine Fitting and Erection; description of Hand and Machine Processes; Turning and Screw Cutting; principles of Constructing and details of Making and Erecting Steam Engines, and the various details of setting out work, etc., etc. By CAMERON KNIGHT, Engineer. *Containing* 1147 *illustrations*, and 397 pages of letter-press. Third edition, 4to, cloth, 18s.

On Designing Belt Gearing. By E. J. COWLING WELCH, Mem. Inst. Mech. Engineers, Author of 'Designing Valve Gearing.' Fcap. 8vo, sewed, 6*d.*

A Handbook of Formulæ, Tables, and Memoranda, for Architectural Surveyors and others engaged in Building. By J. T. HURST, C.E. Thirteenth edition, royal 32mo, roan, 5*s.*

"It is no disparagement to the many excellent publications we refer to, to say that in our opinion this little pocket-book of Hurst's is the very best of them all, without any exception. It would be useless to attempt a recapitulation of the contents, for it appears to contain almost *everything* that *anyone* connected with building could require, and, best of all, made up in a compact form for carrying in the pocket, measuring only 5 in. by 3 in., and about 1 in. thick, in a limp cover. We congratulate the author on the success of his laborious and practically compiled little book, which has received unqualified and deserved praise from every professional person to whom we have shown it."—*The Dublin Builder.*

Tabulated Weights of Angle, Tee, Bulb, Round, Square, and Flat Iron and Steel, and other information for the use of Naval Architects and Shipbuilders. By C. H. JORDAN, M.I.N.A. Fourth edition, 32mo, cloth, 2*s.* 6*d.*

Quantity Surveying. By J. LEANING. With 42 illustrations, crown 8vo, cloth, 9*s.*

CONTENTS:

A complete Explanation of the London Practice.
General Instructions.
Order of Taking Off.
Modes of Measurement of the various Trades.
Use and Waste.
Ventilation and Warming.
Credits, with various Examples of Treatment.
Abbreviations.
Squaring the Dimensions.
Abstracting, with Examples in illustration of each Trade.
Billing.
Examples of Preambles to each Trade.
Form for a Bill of Quantities.
  Do.   Bill of Credits.
  Do.   Bill for Alternative Estimate.
Restorations and Repairs, and Form of Bill.
Variations before Acceptance of Tender.
Errors in a Builder's Estimate.
Schedule of Prices.
Form of Schedule of Prices.
Analysis of Schedule of Prices.
Adjustment of Accounts.
Form of a Bill of Variations.
Remarks on Specifications.
Prices and Valuation of Work, with Examples and Remarks upon each Trade.
The Law as it affects Quantity Surveyors, with Law Reports.
Taking Off after the Old Method.
Northern Practice.
The General Statement of the Methods recommended by the Manchester Society of Architects for taking Quantities.
Examples of Collections.
Examples of "Taking Off" in each Trade.
Remarks on the Past and Present Methods of Estimating.

A Practical Treatise on Heat, as applied to the Useful Arts; for the Use of Engineers, Architects, &c. By THOMAS BOX. With 14 *plates.* Third edition, crown 8vo, cloth, 12*s.* 6*d.*

A Descriptive Treatise on Mathematical Drawing Instruments: their construction, uses, qualities, selection, preservation, and suggestions for improvements, with hints upon Drawing and Colouring. By W. F. STANLEY, M.R.I. Fifth edition, *with numerous illustrations,* crown 8vo, cloth, 5*s.*

Spons' Architects' and Builders' Pocket-Book of Prices and *Memoranda*. Edited by W. YOUNG, Architect. Royal 32mo, roan, 4s. 6d.; or cloth, red edges, 3s. 6d. *Published annually.* Eleventh edition. Now ready.

Long-Span Railway Bridges, comprising Investigations of the Comparative Theoretical and Practical Advantages of the various adopted or proposed Type Systems of Construction, with numerous Formulæ and Tables giving the weight of Iron or Steel required in Bridges from 300 feet to the limiting Spans; to which are added similar Investigations and Tables relating to Short-span Railway Bridges. Second and revised edition. By B. BAKER, Assoc. Inst. C.E. *Plates*, crown 8vo, cloth, 5s.

Elementary Theory and Calculation of Iron Bridges and Roofs. By AUGUST RITTER, Ph.D., Professor at the Polytechnic School at Aix-la-Chapelle. Translated from the third German edition, by H. R. SANKEY, Capt. R.E. With 500 *illustrations*, 8vo, cloth, 15s.

The Builder's Clerk: a Guide to the Management of a Builder's Business. By THOMAS BALES. Fcap. 8vo, cloth, 1s. 6d.

The Elementary Principles of Carpentry. By THOMAS TREDGOLD. Revised from the original edition, and partly re-written, by JOHN THOMAS HURST. Contained in 517 pages of letter-press, and *illustrated with* 48 *plates and* 150 *wood engravings*. Third edition, crown 8vo, cloth, 18s.

> Section I. On the Equality and Distribution of Forces—Section II. Resistance of Timber—Section III. Construction of Floors—Section IV. Construction of Roofs—Section V. Construction of Domes and Cupolas—Section VI. Construction of Partitions—Section VII. Scaffolds, Staging, and Gantries—Section VIII. Construction of Centres for Bridges—Section IX. Coffer-dams, Shoring, and Strutting—Section X. Wooden Bridges and Viaducts—Section XI. Joints, Straps, and other Fastenings—Section XII. Timber.

Our Factories, Workshops, and Warehouses: their Sanitary and Fire-Resisting Arrangements. By B. H. THWAITE, Assoc. Mem. Inst. C.E. With 183 *wood engravings*, crown 8vo, cloth, 9s.

Gold: Its Occurrence and Extraction, embracing the Geographical and Geological Distribution and the Mineralogical Characters of Gold-bearing rocks; the peculiar features and modes of working Shallow Placers, Rivers, and Deep Leads; Hydraulicing; the Reduction and Separation of Auriferous Quartz; the treatment of complex Auriferous ores containing other metals; a Bibliography of the subject and a Glossary of Technical and Foreign Terms. By ALFRED G. LOCK, F.R.G.S. *With numerous illustrations and maps*, 1250 pp., super-royal 8vo, cloth, 2l. 12s. 6d.

A Practical Treatise on Coal Mining. By GEORGE G. ANDRÉ, F.G.S., Assoc. Inst. C.E., Member of the Society of Engineers. With 82 *lithographic plates.* 2 vols., royal 4to, cloth, 3*l*. 12*s*.

Iron Roofs: Examples of Design, Description. *Illustrated with* 64 *Working Drawings of Executed Roofs.* By ARTHUR T. WALMISLEY, Assoc. Mem. Inst. C.E. Imp. 4to, half-morocco, £2 12*s*. 6*d*.

A History of Electric Telegraphy, to the Year 1837. Chiefly compiled from Original Sources, and hitherto Unpublished Documents, by J. J. FAHIE, Mem. Soc. of Tel. Engineers, and of the International Society of Electricians, Paris. Crown 8vo, cloth, 9*s*.

Spons' Information for Colonial Engineers. Edited by J. T. HURST. Demy 8vo, sewed.

No. 1, Ceylon. By ABRAHAM DEANE, C.E. 2*s*. 6*d*.

CONTENTS:

Introductory Remarks—Natural Productions—Architecture and Engineering—Topography, Trade, and Natural History—Principal Stations—Weights and Measures, etc., etc.

No. 2. Southern Africa, including the Cape Colony, Natal, and the Dutch Republics. By HENRY HALL, F.R.G.S., F.R.C.I. With Map. 3*s*. 6*d*.

CONTENTS:

General Description of South Africa—Physical Geography with reference to Engineering Operations—Notes on Labour and Material in Cape Colony—Geological Notes on Rock Formation in South Africa—Engineering Instruments for Use in South Africa—Principal Public Works in Cape Colony: Railways, Mountain Roads and Passes, Harbour Works, Bridges, Gas Works, Irrigation and Water Supply, Lighthouses, Drainage and Sanitary Engineering, Public Buildings, Mines—Table of Woods in South Africa—Animals used for Draught Purposes—Statistical Notes—Table of Distances—Rates of Carriage, etc.

No. 3. India. By F. C. DANVERS, Assoc. Inst. C.E. With Map. 4*s*. 6*d*.

CONTENTS:

Physical Geography of India—Building Materials—Roads—Railways—Bridges—Irrigation—River Works—Harbours—Lighthouse Buildings—Native Labour—The Principal Trees of India—Money—Weights and Measures—Glossary of Indian Terms, etc.

A Practical Treatise on Casting and Founding, including descriptions of the modern machinery employed in the art. By N. E. SPRETSON, Engineer. Third edition, with 82 *plates* drawn to scale, 412 pp., demy 8vo, cloth, 18*s*.

Steam Heating for Buildings; or, Hints to Steam Fitters, being a description of Steam Heating Apparatus for Warming and Ventilating Private Houses and Large Buildings, with remarks on Steam, Water, and Air in their relation to Heating. By W. J. BALDWIN. *With many illustrations.* Fourth edition, crown 8vo, cloth, 10*s*. 6*d*.

The Depreciation of Factories and their Valuation. By EWING MATHESON, M. Inst. C.E. 8vo, cloth, 6s.

A Handbook of Electrical Testing. By H. R. KEMPE, M.S.T.E. Third edition, revised and enlarged, crown 8vo, cloth, 15s.

Gas Works: their Arrangement, Construction, Plant, and Machinery. By F. COLYER, M. Inst. C.E. With 31 *folding plates*, 8vo, cloth, 24s.

The Clerk of Works: a Vade-Mecum for all engaged in the Superintendence of Building Operations. By G. G. HOSKINS, F.R.I.B.A. Third edition, fcap. 8vo, cloth, 1s. 6d.

American Foundry Practice: Treating of Loam, Dry Sand, and Green Sand Moulding, and containing a Practical Treatise upon the Management of Cupolas, and the Melting of Iron. By T. D. WEST, Practical Iron Moulder and Foundry Foreman. Second edition, *with numerous illustrations*, crown 8vo, cloth, 10s. 6d.

The Maintenance of Macadamised Roads. By T. CODRINGTON, M.I.C.E, F.G.S., General Superintendent of County Roads for South Wales. 8vo, cloth, 6s.

Hydraulic Steam and Hand Power Lifting and Pressing Machinery. By FREDERICK COLYER, M. Inst. C.E., M. Inst. M.E. With 73 *plates*, 8vo, cloth, 18s.

Pumps and Pumping Machinery. By F. COLYER, M.I.C.E., M.I.M.E. With 23 *folding plates*, 8vo, cloth, 12s. 6d.

The Municipal and Sanitary Engineer's Handbook. By H. PERCY BOULNOIS, Mem. Inst. C.E., Borough Engineer, Portsmouth. With *numerous illustrations*, demy 8vo, cloth, 12s. 6d.

CONTENTS:

The Appointment and Duties of the Town Surveyor—Traffic—Macadamised Roadways—Steam Rolling—Road Metal and Breaking—Pitched Pavements—Asphalte—Wood Pavements—Footpaths—Kerbs and Gutters—Street Naming and Numbering—Street Lighting—Sewerage—Ventilation of Sewers—Disposal of Sewage—House Drainage—Disinfection—Gas and Water Companies, &c., Breaking up Streets—Improvement of Private Streets—Borrowing Powers—Artizans' and Labourers' Dwellings—Public Conveniences—Scavenging, including Street Cleansing—Watering and the Removing of Snow—Planting Street Trees—Deposit of Plans—Dangerous Buildings—Hoardings—Obstructions—Improving Street Lines—Cellar Openings—Public Pleasure Grounds—Cemeteries—Mortuaries—Cattle and Ordinary Markets—Public Slaughter-houses, etc.—Giving numerous Forms of Notices, Specifications, and General Information upon these and other subjects of great importance to Municipal Engineers and others engaged in Sanitary Work.

Tables of the Principal Speeds occurring in Mechanical Engineering, expressed in metres in a second. By P. KEERAYEFF, Chief Mechanic of the Obouchoff Steel Works, St. Petersburg; translated by SERGIUS KERN, M.E. Fcap. 8vo, sewed, 6*d*.

A Treatise on the Origin, Progress, Prevention, and Cure of Dry Rot in Timber; with Remarks on the Means of Preserving Wood from Destruction by Sea-Worms, Beetles, Ants, etc. By THOMAS ALLEN BRITTON, late Surveyor to the Metropolitan Board of Works, etc., etc. *With* 10 *plates*, crown 8vo, cloth, 7*s*. 6*d*.

Metrical Tables. By G. L. MOLESWORTH, M.I.C.E. 32mo, cloth, 1*s*. 6*d*.

CONTENTS.

General—Linear Measures—Square Measures—Cubic Measures—Measures of Capacity—Weights—Combinations—Thermometers.

Elements of Construction for Electro-Magnets. By Count TH. DU MONCEL, Mem. de l'Institut de France. Translated from the French by C. J. WHARTON. Crown 8vo, cloth, 4*s*. 6*d*.

Electro-Telegraphy. By FREDERICK S. BEECHEY, Telegraph Engineer. A Book for Beginners. *Illustrated.* Fcap. 8vo, sewed, 6*d*.

Handrailing: by the Square Cut. By JOHN JONES, Staircase Builder. Part Second, *with eight plates*, 8vo, cloth, 3*s*. 6*d*.

Practical Electrical Units Popularly Explained, with *numerous illustrations* and Remarks. By JAMES SWINBURNE, late of J. W. Swan and Co., Paris, late of Brush-Swan Electric Light Company, U.S.A. 18mo, cloth, 1*s*. 6*d*.

Philipp Reis, Inventor of the Telephone: A Biographical Sketch. With Documentary Testimony, Translations of the Original Papers of the Inventor, &c. By SILVANUS P. THOMPSON, B.A., Dr. Sc., Professor of Experimental Physics in University College, Bristol. *With illustrations*, 8vo, cloth, 7*s*. 6*d*.

A Treatise on the Use of Belting for the Transmission of Power. By J. H. COOPER. Second edition, *illustrated*, 8vo, cloth, 15*s*.

A Pocket-Book of Useful Formulæ and Memoranda for Civil and Mechanical Engineers. By GUILFORD L. MOLESWORTH, Mem. Inst. C.E., Consulting Engineer to the Government of India for State Railways. *With numerous illustrations*, 744 pp. Twenty-first edition, revised and enlarged, 32mo, roan, 6s.

SYNOPSIS OF CONTENTS:

Surveying, Levelling, etc.—Strength and Weight of Materials—Earthwork, Brickwork, Masonry, Arches, etc.—Struts, Columns, Beams, and Trusses—Flooring, Roofing, and Roof Trusses—Girders, Bridges, etc.—Railways and Roads—Hydraulic Formulæ—Canals, Sewers, Waterworks, Docks—Irrigation and Breakwaters—Gas, Ventilation, and Warming—Heat, Light, Colour, and Sound—Gravity: Centres, Forces, and Powers—Millwork, Teeth of Wheels, Shafting, etc.—Workshop Recipes—Sundry Machinery—Animal Power—Steam and the Steam Engine—Water-power, Water-wheels, Turbines, etc.—Wind and Windmills—Steam Navigation, Ship Building, Tonnage, etc.—Gunnery, Projectiles, etc.—Weights, Measures, and Money—Trigonometry, Conic Sections, and Curves—Telegraphy—Mensuration—Tables of Areas and Circumference, and Arcs of Circles—Logarithms, Square and Cube Roots, Powers—Reciprocals, etc.—Useful Numbers—Differential and Integral Calculus—Algebraic Signs—Telegraphic Construction and Formulæ.

Spons' Tables and Memoranda for Engineers; selected and arranged by J. T. HURST, C.E., Author of 'Architectural Surveyors' Handbook,' 'Hurst's Tredgold's Carpentry,' etc. Seventh edition, 64mo, roan, gilt edges, 1s.; or in cloth case, 1s. 6d.

This work is printed in a pearl type, and is so small, measuring only 2½ in. by 1¾ in. by ⅜ in. thick, that it may be easily carried in the waistcoat pocket.

"It is certainly an extremely rare thing for a reviewer to be called upon to notice a volume measuring but 2½ in. by 1¾ in., yet these dimensions faithfully represent the size of the handy little book before us. The volume—which contains 118 printed pages, besides a few blank pages for memoranda—is, in fact, a true pocket-book, adapted for being carried in the waistcoat pocket, and containing a far greater amount and variety of information than most people would imagine could be compressed into so small a space. . . . The little volume has been compiled with considerable care and judgment, and we can cordially recommend it to our readers as a useful little pocket companion."—*Engineering.*

A Practical Treatise on Natural and Artificial Concrete, its Varieties and Constructive Adaptations. By HENRY REID, Author of the 'Science and Art of the Manufacture of Portland Cement.' New Edition, *with 59 woodcuts and 5 plates*, 8vo, cloth, 15s.

Hydrodynamics: Treatise relative to the Testing of Water-Wheels and Machinery, with various other matters pertaining to Hydrodynamics. By JAMES EMERSON. *With numerous illustrations*, 360 pp. Third edition, crown 8vo, cloth, 4s. 6d.

Electricity as a Motive Power. By Count TH. DU MONCEL, Membre de l'Institut de France, and FRANK GERALDY, Ingénieur des Ponts et Chaussées. Translated and Edited, with Additions, by C. J. WHARTON, Assoc. Soc. Tel. Eng. and Elec. *With 113 engravings and diagrams*, crown 8vo, cloth, 7s. 6d.

Hints on Architectural Draughtsmanship. By G. W. TUXFORD HALLATT. Fcap. 8vo, cloth, 1s. 6d.

Treatise on Valve-Gears, with special consideration of the Link-Motions of Locomotive Engines. By Dr. GUSTAV ZEUNER, Professor of Applied Mechanics at the Confederated Polytechnikum of Zurich. Translated from the Fourth German Edition, by Professor J. F. KLEIN, Lehigh University, Bethlehem, Pa. *Illustrated*, 8vo, cloth, 12s. 6d.

The French-Polisher's Manual. By a French-Polisher; containing Timber Staining, Washing, Matching, Improving, Painting, Imitations, Directions for Staining, Sizing, Embodying, Smoothing, Spirit Varnishing, French-Polishing, Directions for Re-polishing. Third edition, royal 32mo, sewed, 6d.

Hops, their Cultivation, Commerce, and Uses in various Countries. By P. L. SIMMONDS. Crown 8vo, cloth, 4s. 6d.

A Practical Treatise on the Manufacture and Distri-bution of Coal Gas. By WILLIAM RICHARDS. Demy 4to, with *numerous wood engravings and 29 plates*, cloth, 28s.

SYNOPSIS OF CONTENTS:

Introduction—History of Gas Lighting—Chemistry of Gas Manufacture, by Lewis Thompson, Esq., M.R.C.S.—Coal, with Analyses, by J. Paterson, Lewis Thompson, and G. R. Hislop, Esqrs.—Retorts, Iron and Clay—Retort Setting—Hydraulic Main—Condensers—Exhausters—Washers and Scrubbers—Purifiers—Purification—History of Gas Holder—Tanks, Brick and Stone, Composite, Concrete, Cast-iron, Compound Annular Wrought-iron—Specifications—Gas Holders—Station Meter—Governor—Distribution—Mains—Gas Mathematics, or Formulæ for the Distribution of Gas, by Lewis Thompson, Esq.—Services—Consumers' Meters—Regulators—Burners—Fittings—Photometer—Carburization of Gas—Air Gas and Water Gas—Composition of Coal Gas, by Lewis Thompson, Esq.—Analyses of Gas—Influence of Atmospheric Pressure and Temperature on Gas—Residual Products—Appendix—Description of Retort Settings, Buildings, etc., etc.

Practical Geometry, Perspective, and Engineering Drawing; a Course of Descriptive Geometry adapted to the Requirements of the Engineering Draughtsman, including the determination of cast shadows and Isometric Projection, each chapter being followed by numerous examples; to which are added rules for Shading, Shade-lining, etc., together with practical instructions as to the Lining, Colouring, Printing, and general treatment of Engineering Drawings, with a chapter on drawing Instruments. By GEORGE S. CLARKE, Capt. R.E. Second edition, *with 21 plates*. 2 vols., cloth, 10s. 6d.

The Elements of Graphic Statics. By Professor KARL VON OTT, translated from the German by G. S. CLARKE, Capt. R.E., Instructor in Mechanical Drawing, Royal Indian Engineering College. *With 93 illustrations*, crown 8vo, cloth, 5s.

The Principles of Graphic Statics. By GEORGE SYDENHAM CLARKE, Capt. Royal Engineers. *With 112 illustrations.* 4to, cloth, 12s. 6d.

Dynamo-Electric Machinery: A Manual for Students of Electro-technics. By SILVANUS P. THOMPSON, B.A., D.Sc., Professor of Experimental Physics in University College, Bristol, etc., etc. Second edition, *illustrated*, 8vo, cloth, 12s. 6d.

The New Formula for Mean Velocity of Discharge of Rivers and Canals. By W. R. KUTTER. Translated from articles in the 'Cultur-Ingénieur,' by LOWIS D'A. JACKSON, Assoc. Inst. C.E. 8vo, cloth, 12s. 6d.

Practical Hydraulics; a Series of Rules and Tables for the use of Engineers, etc., etc. By THOMAS BOX. Fifth edition, numerous plates, post 8vo, cloth, 5s.

A Practical Treatise on the Construction of Horizontal and Vertical Waterwheels, specially designed for the use of operative mechanics. By WILLIAM CULLEN, Millwright and Engineer. With 11 plates. Second edition, revised and enlarged, small 4to, cloth, 12s. 6d.

Tin: Describing the Chief Methods of Mining, Dressing and Smelting it abroad; with Notes upon Arsenic, Bismuth and Wolfram. By ARTHUR G. CHARLETON, Mem. American Inst. of Mining Engineers. With plates, 8vo, cloth, 12s. 6d.

Perspective, Explained and Illustrated. By G. S. CLARKE, Capt. R.E. With illustrations, 8vo, cloth, 3s. 6d.

The Essential Elements of Practical Mechanics; based on the Principle of Work, designed for Engineering Students. By OLIVER BYRNE, formerly Professor of Mathematics, College for Civil Engineers. Third edition, with 148 wood engravings, post 8vo, cloth, 7s. 6d.

CONTENTS:

Chap. 1. How Work is Measured by a Unit, both with and without reference to a Unit of Time—Chap. 2. The Work of Living Agents, the Influence of Friction, and introduces one of the most beautiful Laws of Motion—Chap. 3. The principles expounded in the first and second chapters are applied to the Motion of Bodies—Chap. 4. The Transmission of Work by simple Machines—Chap. 5. Useful Propositions and Rules.

The Practical Millwright and Engineer's Ready Reckoner; or Tables for finding the diameter and power of cog-wheels, diameter, weight, and power of shafts, diameter and strength of bolts, etc. By THOMAS DIXON. Fourth edition, 12mo, cloth, 3s.

Breweries and Maltings: their Arrangement, Construction, Machinery, and Plant. By G. SCAMELL, F.R.I.B.A. Second edition, revised, enlarged, and partly rewritten. By F. COLYER, M.I.C.E., M.I.M.E. With 20 plates, 8vo, cloth, 18s.

A Practical Treatise on the Manufacture of Starch, Glucose, Starch-Sugar, and Dextrine, based on the German of L. Von Wagner, Professor in the Royal Technical School, Buda Pesth, and other authorities. By JULIUS FRANKEL; edited by ROBERT HUTTER, proprietor of the Philadelphia Starch Works. With 58 illustrations, 344 pp., 8vo, cloth, 18s.

A Practical Treatise on Mill-gearing, Wheels, Shafts,
Riggers, etc.; for the use of Engineers. By THOMAS BOX. Third edition, *with* 11 *plates.* Crown 8vo, cloth, 7s. 6d.

Mining Machinery: a Descriptive Treatise on the Machinery, Tools, and other Appliances used in Mining. By G. G. ANDRÉ, F.G.S., Assoc. Inst. C.E., Mem. of the Society of Engineers. Royal 4to, uniform with the Author's Treatise on Coal Mining, containing 182 *plates*, accurately drawn to scale, with descriptive text, in 2 vols., cloth, 3*l.* 12*s.*

CONTENTS:

Machinery for Prospecting, Excavating, Hauling, and Hoisting—Ventilation—Pumping—Treatment of Mineral Products, including Gold and Silver, Copper, Tin, and Lead, Iron Coal, Sulphur, China Clay, Brick Earth, etc.

Tables for Setting out Curves for Railways, Canals,
Roads, etc., varying from a radius of five chains to three miles. By A. KENNEDY and R. W. HACKWOOD. *Illustrated*, 32mo, cloth, 2*s.* 6*d.*

The Science and Art of the Manufacture of Portland
Cement, with observations on some of its constructive applications. With 66 *illustrations.* By HENRY REID, C.E., Author of 'A Practical Treatise on Concrete,' etc., etc. 8vo, cloth, 18*s.*

The Draughtsman's Handbook of Plan and Map
Drawing; including instructions for the preparation of Engineering, Architectural, and Mechanical Drawings. *With numerous illustrations in the text, and* 33 *plates* (15 *printed in colours*). By G. G. ANDRÉ, F.G.S., Assoc. Inst. C.E. 4to, cloth, 9*s.*

CONTENTS:

The Drawing Office and its Furnishings—Geometrical Problems—Lines, Dots, and their Combinations—Colours, Shading, Lettering, Bordering, and North Points—Scales—Plotting—Civil Engineers' and Surveyors' Plans—Map Drawing—Mechanical and Architectural Drawing—Copying and Reducing Trigonometrical Formulæ, etc., etc.

The Boiler-maker's and Iron Ship-builder's Companion,
comprising a series of original and carefully calculated tables, of the utmost utility to persons interested in the iron trades. By JAMES FODEN, author of 'Mechanical Tables,' etc. Second edition revised, *with illustrations*, crown 8vo, cloth, 5*s.*

Rock Blasting: a Practical Treatise on the means employed in Blasting Rocks for Industrial Purposes. By G. G. ANDRÉ, F.G.S., Assoc. Inst. C.E. *With* 56 *illustrations and* 12 *plates*, 8vo, cloth, 10*s.* 6*d.*

Painting and Painters' Manual: a Book of Facts for Painters and those who Use or Deal in Paint Materials. By C. L. CONDIT and J. SCHELLER. *Illustrated*, 8vo, cloth, 10*s.* 6*d.*

A Treatise on Ropemaking as practised in public and private Rope-yards, with a Description of the Manufacture, Rules, Tables of Weights, etc., adapted to the Trade, Shipping, Mining, Railways, Builders, etc. By R. CHAPMAN, formerly foreman to Messrs. Huddart and Co., Limehouse, and late Master Ropemaker to H.M. Dockyard, Deptford. Second edition, 12mo, cloth, 3s.

Laxton's Builders' and Contractors' Tables; for the use of Engineers, Architects, Surveyors, Builders, Land Agents, and others. Bricklayer, containing 22 tables, with nearly 30,000 calculations. 4to, cloth, 5s.

Laxton's Builders' and Contractors' Tables. Excavator, Earth, Land, Water, and Gas, containing 53 tables, with nearly 24,000 calculations. 4to, cloth, 5s.

Sanitary Engineering: a Guide to the Construction of Works of Sewerage and House Drainage, with Tables for facilitating the calculations of the Engineer. By BALDWIN LATHAM, C.E., M. Inst. C.E., F.G.S., F.M.S., Past-President of the Society of Engineers. Second edition, *with numerous plates and woodcuts*, 8vo, cloth, 1l. 10s.

Screw Cutting Tables for Engineers and Machinists, giving the values of the different trains of Wheels required to produce Screws of any pitch, calculated by Lord Lindsay, M.P., F.R.S., F.R.A.S., etc. Cloth, oblong, 2s.

Screw Cutting Tables, for the use of Mechanical Engineers, showing the proper arrangement of Wheels for cutting the Threads of Screws of any required pitch, with a Table for making the Universal Gas-pipe Threads and Taps. By W. A. MARTIN, Engineer. Second edition, oblong, cloth, 1s., or sewed, 6d.

A Treatise on a Practical Method of Designing Slide-Valve Gears by Simple Geometrical Construction, based upon the principles enunciated in Euclid's Elements, and comprising the various forms of Plain Slide-Valve and Expansion Gearing; together with Stephenson's, Gooch's, and Allan's Link-Motions, as applied either to reversing or to variable expansion combinations. By EDWARD J. COWLING WELCH, Memb. Inst. Mechanical Engineers. Crown 8vo, cloth, 6s.

Cleaning and Scouring: a Manual for Dyers, Laundresses, and for Domestic Use. By S. CHRISTOPHER. 18mo, sewed, 6d.

A Handbook of House Sanitation; for the use of all persons seeking a Healthy Home. A reprint of those portions of Mr. Bailey-Denton's Lectures on Sanitary Engineering, given before the School of Military Engineering, which related to the "Dwelling," enlarged and revised by his Son, E. F. BAILEY-DENTON, C.E., B.A. *With* 140 *illustrations*, 8vo, cloth, 8s. 6d.

A Glossary of Terms used in Coal Mining. By
WILLIAM STUKELEY GRESLEY, Assoc. Mem. Inst. C.E., F.G.S., Member
of the North of England Institute of Mining Engineers. *Illustrated with
numerous woodcuts and diagrams,* crown 8vo, cloth, 5s.

A Pocket-Book for Boiler Makers and Steam Users,
comprising a variety of useful information for Employer and Workman,
Government Inspectors, Board of Trade Surveyors, Engineers in charge
of Works and Slips, Foremen of Manufactories, and the general Steam-
using Public. By MAURICE JOHN SEXTON. Second edition, royal
32mo, roan, gilt edges, 5s.

Electrolysis: a Practical Treatise on Nickeling,
Coppering, Gilding, Silvering, the Refining of Metals, and the treatment
of Ores by means of Electricity. By HIPPOLYTE FONTAINE, translated
from the French by J. A. BERLY, C.E., Assoc. S.T.E. *With engravings.*
8vo, cloth, 9s.

A Practical Treatise on the Steam Engine, containing Plans and Arrangements of Details for Fixed Steam Engines,
with Essays on the Principles involved in Design and Construction. By
ARTHUR RIGG, Engineer, Member of the Society of Engineers and of
the Royal Institution of Great Britain. Demy 4to, *copiously illustrated
with woodcuts and 96 plates,* in one Volume, half-bound morocco, 2*l.* 2s.;
or cheaper edition, cloth, 25s.

> This work is not, in any sense, an elementary treatise, or history of the steam engine, but
> is intended to describe examples of Fixed Steam Engines without entering into the wide
> domain of locomotive or marine practice. To this end illustrations will be given of the most
> recent arrangements of Horizontal, Vertical, Beam, Pumping, Winding, Portable, Semi-
> portable, Corliss, Allen, Compound, and other similar Engines, by the most eminent Firms in
> Great Britain and America. The laws relating to the action and precautions to be observed
> in the construction of the various details, such as Cylinders, Pistons, Piston-rods, Connecting-
> rods, Cross-heads, Motion-blocks, Eccentrics, Simple, Expansion, Balanced, and Equilibrium
> Slide-valves, and Valve-gearing will be minutely dealt with. In this connection will be found
> articles upon the Velocity of Reciprocating Parts and the Mode of Applying the Indicator,
> Heat and Expansion of Steam Governors, and the like. It is the writer's desire to draw
> illustrations from every possible source, and give only those rules that present practice deems
> correct.

Barlow's Tables of Squares, Cubes, Square Roots,
Cube Roots, Reciprocals of all Integer Numbers up to 10,000. Post 8vo,
cloth, 6s.

Camus (M.) Treatise on the Teeth of Wheels, demon-
strating the best forms which can be given to them for the purposes of
Machinery, such as Mill-work and Clock-work, and the art of finding
their numbers. Translated from the French, with details of the present
practice of Millwrights, Engine Makers, and other Machinists, by
ISAAC HAWKINS. Third edition, *with 18 plates,* 8vo, cloth, 5s.

A Practical Treatise on the Science of Land and Engineering Surveying, Levelling, Estimating Quantities, etc., with a general description of the several Instruments required for Surveying, Levelling, Plotting, etc. By H. S. MERRETT. Fourth edition, revised by G. W. USILL, Assoc. Mem. Inst. C.E. 41 *plates, with illustrations and tables,* royal 8vo, cloth, 12s. 6d.

PRINCIPAL CONTENTS:

Part 1. Introduction and the Principles of Geometry. Part 2. Land Surveying: comprising General Observations—The Chain—Offsets Surveying by the Chain only—Surveying Hilly Ground—To Survey an Estate or Parish by the Chain only—Surveying with the Theodolite—Mining and Town Surveying—Railroad Surveying—Mapping—Division and Laying out of Land—Observations on Enclosures—Plane Trigonometry. Part 3. Levelling—Simple and Compound Levelling—The Level Book—Parliamentary Plan and Section—Levelling with a Theodolite—Gradients—Wooden Curves—To Lay out a Railway Curve—Setting out Widths. Part 4. Calculating Quantities generally for Estimates—Cuttings and Embankments—Tunnels—Brickwork—Ironwork—Timber Measuring. Part 5. Description and Use of Instruments in Surveying and Plotting—The Improved Dumpy Level—Troughton's Level—The Prismatic Compass—Proportional Compass—Box Sextant—Vernier—Pantagraph—Merrett's Improved Quadrant—Improved Computation Scale—The Diagonal Scale—Straight Edge and Sector. Part 6. Logarithms of Numbers—Logarithmic Sines and Co-Sines, Tangents and Co-Tangents—Natural Sines and Co-Sines—Tables for Earthwork, for Setting out Curves, and for various Calculations, etc., etc., etc.

Saws: the History, Development, Action, Classification, and Comparison of Saws of all kinds. By ROBERT GRIMSHAW. With 220 *illustrations,* 4to, cloth, 12s. 6d.

A Supplement to the above; containing additional practical matter, more especially relating to the forms of Saw Teeth for special material and conditions, and to the behaviour of Saws under particular conditions. *With* 120 *illustrations,* cloth, 9s.

A Guide for the Electric Testing of Telegraph Cables. By Capt. V. HOSKIŒR, Royal Danish Engineers. *With illustrations,* second edition, crown 8vo, cloth, 4s. 6d.

Laying and Repairing Electric Telegraph Cables. By Capt. V. HOSKIŒR, Royal Danish Engineers. Crown 8vo, cloth, 3s. 6d.

A Pocket-Book of Practical Rules for the Proportions of Modern Engines and Boilers for Land and Marine purposes. By N. P. BURGH. Seventh edition, royal 32mo, roan, 4s. 6d.

The Assayer's Manual: an Abridged Treatise on the Docimastic Examination of Ores and Furnace and other Artificial Products. By BRUNO KERL. Translated by W. T. BRANNT. *With* 65 *illustrations,* 8vo, cloth, 12s. 6d.

The Steam Engine considered as a Heat Engine: a Treatise on the Theory of the Steam Engine, illustrated by Diagrams, Tables, and Examples from Practice. By JAS. H. COTTERILL, M.A., F.R.S., Professor of Applied Mechanics in the Royal Naval College. 8vo, cloth, 12s. 6d.

Electricity: its Theory, Sources, and Ap lications.
By J. T. SPRAGUE, M.S.T.E. Second edition, revised and enlarged, *with numerous illustrations*, crown 8vo, cloth, 15s.

The Practice of Hand Turning in Wood, Ivory, Shell, etc., with Instructions for Turning such Work in Metal as may be required in the Practice of Turning in Wood, Ivory, etc.; also an Appendix on Ornamental Turning. (A book for beginners.) By FRANCIS CAMPIN. Third edition, *with wood engravings*, crown 8vo, cloth, 6s.

CONTENTS:

On Lathes—Turning Tools—Turning Wood—Drilling—Screw Cutting—Miscellaneous Apparatus and Processes—Turning Particular Forms—Staining—Polishing—Spinning Metals—Materials—Ornamental Turning, etc.

Health and Comfort in House Building, or Ventilation with Warm Air by Self-Acting Suction Power, with Review of the mode of Calculating the Draught in Hot-Air Flues, and with some actual Experiments. By J. DRYSDALE, M.D., and J. W. HAYWARD, M.D. Second edition, with Supplement, *with plates*, demy 8vo, cloth, 7s. 6d.

Treatise on Watchwork, Past and Present. By the Rev. H. L. NELTHROPP, M.A., F.S.A. *With 32 illustrations*, crown 8vo, cloth, 6s. 6d.

CONTENTS:

Definitions of Words and Terms used in Watchwork—Tools—Time—Historical Summary—On Calculations of the Numbers for Wheels and Pinions; their Proportional Sizes, Trains, etc.—Of Dial Wheels, or Motion Work—Length of Time of Going without Winding up—The Verge—The Horizontal—The Duplex—The Lever—The Chronometer—Repeating Watches—Keyless Watches—The Pendulum, or Spiral Spring—Compensation—Jewelling of Pivot Holes—Clerkenwell—Fallacies of the Trade—Incapacity of Workmen—How to Choose and Use a Watch, etc.

Notes in Mechanical Engineering. Compiled principally for the use of the Students attending the Classes on this subject at the City of London College. By HENRY ADAMS, Mem. Inst. M.E., Mem. Inst. C.E., Mem. Soc. of Engineers. Crown 8vo, cloth, 2s. 6d.

Algebra Self-Taught. By W. P. HIGGS, M.A., D.Sc., LL.D., Assoc. Inst. C.E., Author of 'A Handbook of the Differential Calculus,' etc. Second edition, crown 8vo, cloth, 2s. 6d.

CONTENTS:

Symbols and the Signs of Operation—The Equation and the Unknown Quantity—Positive and Negative Quantities—Multiplication—Involution—Exponents—Negative Exponents—Roots, and the Use of Exponents as Logarithms—Logarithms—Tables of Logarithms and Proportionate Parts—Transformation of System of Logarithms—Common Uses of Common Logarithms—Compound Multiplication and the Binomial Theorem—Division, Fractions, and Ratio—Continued Proportion—The Series and the Summation of the Series—Limit of Series—Square and Cube Roots—Equations—List of Formulæ, etc.

Spons' Dictionary of Engineering, Civil, Mechanical, Military, and Naval; with technical terms in French, German, Italian, and Spanish, 3100 pp., and *nearly* 8000 *engravings*, in super-royal 8vo, in 8 divisions, 5l. 8s. Complete in 3 vols., cloth, 5l. 5s. Bound in a superior manner, half-morocco, top edge gilt, 3 vols., 6l. 12s.

Canoe and Boat Building: a complete Manual for Amateurs, containing plain and comprehensive directions for the construction of Canoes, Rowing and Sailing Boats, and Hunting Craft. By W. P. STEPHENS. *With numerous illustrations and 24 plates of Working Drawings.* Crown 8vo, cloth, 7s. 6d.

Cultural Industries for Queensland: Papers on the cultivation of useful Plants suited to the climate of Queensland, their value as Food, in the Arts, and in Medicine, and methods of obtaining their products. By L. A. BERNAYS, F.L.S., F.R.G.S. 8vo, half calf, 7s. 6d. The same, in cloth, 6s.

Proceedings of the National Conference of Electricians, *Philadelphia,* October 8th to 13th, 1884. 18mo, cloth, 3s.

Dynamo-Electricity, its Generation, Application, Transmission, Storage, and Measurement. By G. B. PRESCOTT. *With 545 illustrations.* 8vo, cloth, 1l. 1s.

Domestic Electricity for Amateurs. Translated from the French of E. HOSPITALIER, Editor of "L'Electricien," by C. J. WHARTON, Assoc. Soc. Tel. Eng. *Numerous illustrations.* Demy 8vo, cloth, 9s.

CONTENTS:

1. Production of the Electric Current—2. Electric Bells—3. Automatic Alarms—4. Domestic Telephones—5. Electric Clocks—6. Electric Lighters—7. Domestic Electric Lighting—8. Domestic Application of the Electric Light—9. Electric Motors—10. Electrical Locomotion—11. Electrotyping, Plating, and Gilding—12. Electric Recreations—13. Various applications—Workshop of the Electrician.

Wrinkles in Electric Lighting. By VINCENT STEPHEN. *With illustrations.* 18mo, cloth, 2s. 6d.

CONTENTS:

1. The Electric Current and its production by Chemical means—2. Production of Electric Currents by Mechanical means—3. Dynamo-Electric Machines—4. Electric Lamps—5. Lead—6. Ship Lighting.

The Practical Flax Spinner; being a Description of the Growth, Manipulation, and Spinning of Flax and Tow. By LESLIE C. MARSHALL, of Belfast. *With illustrations.* 8vo, cloth, 15s.

Foundations and Foundation Walls for all classes of Buildings, Pile Driving, Building Stones and Bricks, Pier and Wall construction, Mortars, Limes, Cements, Concretes, Stuccos, &c. 64 *illustrations.* By G. T. POWELL and F. BAUMAN. 8vo, cloth, 10s. 6d.

The British Jugernath. Free Trade! Fair Trade!! Reciprocity!!! and Retaliation!!!! By G. L. M. 8vo, sewed. 6d.

Manual for Gas Engineering Students. By D. Lee. 18mo, cloth 1s.

Hydraulic Machinery, Past and Present. A Lecture delivered to the London and Suburban Railway Officials' Association. By H. Adams, Mem. Inst. C.E. *Folding plate.* 8vo, sewed, 1s.

Twenty Years with the Indicator. By Thomas Pray, Jun., C.E., M.E., Member of the American Society of Civil Engineers. 2 vols., royal 8vo, cloth, 12s. 6d.

Annual Statistical Report of the Secretary to the Members of the Iron and Steel Association on the Home and Foreign Iron and Steel Industries in 1884. Issued March 1885. 8vo, sewed, 5s.

Bad Drains, and How to Test them; with Notes on the Ventilation of Sewers, Drains, and Sanitary Fittings, and the Origin and Transmission of Zymotic Disease. By R. Harris Reeves. Crown 8vo, cloth, 3s. 6d.

Standard Practical Plumbing; being a complete Encyclopædia for Practical Plumbers and Guide for Architects, Builders, Gas Fitters, Hot-water Fitters, Ironmongers, Lead Burners, Sanitary Engineers, Zinc Workers, &c. *Illustrated by over 2000 engravings.* By P. J. Davies. Vol. I, royal 8vo, cloth, 7s. 6d.

Pneumatic Transmission of Messages and Parcels between Paris and London, *viâ* Calais and Dover. By J. B. Berlier, C.E. Small folio, sewed, 6d.

List of Tests (Reagents), arranged in alphabetical order, according to the names of the originators. Designed especially for the convenient reference of Chemists, Pharmacists, and Scientists. By Hans M. Wilder. Crown 8vo, cloth, 4s. 6d.

Ten Years' Experience in Works of Intermittent Downward Filtration. By J. Bailey Denton, Mem. Inst. C.E. Second edition, with additions. Royal 8vo, sewed, 4s.

A Treatise on the Manufacture of Soap and Candles, Lubricants and Glycerin. By W. Lant Carpenter, B.A., B.Sc. (late of Messrs. C. Thomas and Brothers, Bristol). *With illustrations.* Crown 8vo, cloth, 10s. 6d.

The Stability of Ships explained simply, and calculated by a new Graphic method. By J. C. SPENCE, M.I.N.A. 4to, sewed, 3s. 6d.

Steam Making, or Boiler Practice. By CHARLES A. SMITH, C.E. 8vo, cloth, 9s.

CONTENTS:

1. The Nature of Heat and the Properties of Steam—2. Combustion.—3. Externally Fired Stationary Boilers—4. Internally Fired Stationary Boilers—5. Internally Fired Portable Locomotive and Marine Boilers—6. Design, Construction, and Strength of Boilers—7. Proportions of Heating Surface, Economic Evaporation, Explosions—8. Miscellaneous Boilers, Choice of Boiler Fittings and Appurtenances.

The Fireman's Guide; a Handbook on the Care of Boilers. By TEKNOLOG, föreningen T. I. Stockholm. Translated from the third edition, and revised by KARL P. DAHLSTROM, M.E. Second edition. Fcap. 8vo, cloth, 2s.

A Treatise on Modern Steam Engines and Boilers, including Land Locomotive, and Marine Engines and Boilers, for the use of Students. By FREDERICK COLYER, M. Inst. C.E., Mem. Inst. M.E. With 36 plates. 4to, cloth. (*Nearly ready.*)

CONTENTS:

1. Introduction—2. Original Engines—3. Boilers—4. High-Pressure Beam Engines—5. Cornish Beam Engines—6. Horizontal Engines—7. Oscillating Engines—8. Vertical High-Pressure Engines—9. Special Engines—10. Portable Engines—11. Locomotive Engines—12. Marine Engines.

Steam Engine Management; a Treatise on the Working and Management of Steam Boilers. By F. COLYER, M. Inst. C.E., Mem. Inst. M.E. 18mo, cloth, 2s.

Land Surveying on the Meridian and Perpendicular System. By WILLIAM PENMAN, C.E. 8vo, cloth, 8s. 6d.

The Topographer, his Instruments and Methods, designed for the use of Students, Amateur Photographers, Surveyors, Engineers, and all persons interested in the location and construction of works based upon Topography. *Illustrated with numerous plates, maps, and engravings.* By LEWIS M. HAUPT, A.M. 8vo, cloth, 18s.

A Text-Book of Tanning, embracing the Preparation of all kinds of Leather. By HARRY R. PROCTOR, F.C.S., of Low Lights Tanneries. *With illustrations.* Crown 8vo, cloth, 10s. 6d.

In super-royal 8vo, 1168 pp., *with* 2400 *illustrations*, in 3 Divisions, cloth, price 13s. 6d. each; or 1 vol., cloth, 2l.; or half-morocco, 2l. 8s.

A SUPPLEMENT

TO

SPONS' DICTIONARY OF ENGINEERING.

EDITED BY ERNEST SPON, MEMB. SOC. ENGINEERS.

Abacus, Counters, Speed Indicators, and Slide Rule.
Agricultural Implements and Machinery.
Air Compressors.
Animal Charcoal Machinery.
Antimony.
Axles and Axle-boxes.
Barn Machinery.
Belts and Belting.
Blasting. Boilers.
Brakes.
Brick Machinery.
Bridges.
Cages for Mines.
Calculus, Differential and Integral.
Canals.
Carpentry.
Cast Iron.
Cement, Concrete, Limes, and Mortar.
Chimney Shafts.
Coal Cleansing and Washing.

Coal Mining.
Coal Cutting Machines.
Coke Ovens. Copper.
Docks. Drainage.
Dredging Machinery.
Dynamo - Electric and Magneto-Electric Machines.
Dynamometers.
Electrical Engineering, Telegraphy, Electric Lighting and its practical details, Telephones
Engines, Varieties of.
Explosives. Fans.
Founding, Moulding and the practical work of the Foundry.
Gas, Manufacture of.
Hammers, Steam and other Power.
Heat. Horse Power.
Hydraulics.
Hydro-geology.
Indicators. Iron.
Lifts, Hoists, and Elevators.

Lighthouses, Buoys, and Beacons.
Machine Tools.
Materials of Construction.
Meters.
Ores, Machinery and Processes employed to Dress.
Piers.
Pile Driving.
Pneumatic Transmission.
Pumps.
Pyrometers.
Road Locomotives.
Rock Drills.
Rolling Stock.
Sanitary Engineering.
Shafting.
Steel.
Steam Navvy.
Stone Machinery.
Tramways.
Well Sinking.

London: E. & F. N. SPON, 125, Strand.
New York: 35, Murray Street.

NOW COMPLETE.

With nearly 1500 *illustrations*, in super-royal 8vo, in 5 Divisions, cloth. Divisions 1 to 4, 13s. 6d. each; Division 5, 17s. 6d.; or 2 vols., cloth, £3 10s.

SPONS' ENCYCLOPÆDIA
OF THE
INDUSTRIAL ARTS, MANUFACTURES, AND COMMERCIAL PRODUCTS.

EDITED BY C. G. WARNFORD LOCK, F.L.S.

Among the more important of the subjects treated of, are the following:—

Acids, 207 pp. 220 figs.
Alcohol, 23 pp. 16 figs.
Alcoholic Liquors, 13 pp.
Alkalies, 89 pp. 78 figs.
Alloys.   Alum.
Asphalt.   Assaying.
Beverages, 89 pp. 29 figs.
Blacks.
Bleaching Powder, 15 pp.
Bleaching, 51 pp. 48 figs.
Candles, 18 pp. 9 figs.
Carbon Bisulphide.
Celluloid, 9 pp.
Cements.   Clay.
Coal-tar Products, 44 pp. 14 figs.
Cocoa, 8 pp.
Coffee, 32 pp. 13 figs.
Cork, 8 pp. 17 figs.
Cotton Manufactures, 62 pp. 57 figs.
Drugs, 38 pp.
Dyeing and Calico Printing, 28 pp. 9 figs.
Dyestuffs, 16 pp.
Electro-Metallurgy, 13 pp.
Explosives, 22 pp. 33 figs.
Feathers.
Fibrous Substances, 92 pp. 79 figs.
Floor-cloth, 16 pp. 21 figs.
Food Preservation, 8 pp.
Fruit, 8 pp.

Fur, 5 pp.
Gas, Coal, 8 pp.
Gems.
Glass, 45 pp. 77 figs.
Graphite, 7 pp.
Hair, 7 pp.
Hair Manufactures.
Hats, 26 pp. 26 figs.
Honey.   Hops.
Horn.
Ice, 10 pp. 14 figs.
Indiarubber Manufactures, 23 pp. 17 figs.
Ink, 17 pp.
Ivory.
Jute Manufactures, 11 pp., 11 figs.
Knitted Fabrics — Hosiery, 15 pp. 13 figs.
Lace, 13 pp. 9 figs.
Leather, 28 pp. 31 figs.
Linen Manufactures, 16 pp. 6 figs.
Manures, 21 pp. 30 figs.
Matches, 17 pp. 38 figs.
Mordants, 13 pp.
Narcotics, 47 pp.
Nuts, 10 pp.
Oils and Fatty Substances, 125 pp.
Paint.
Paper, 26 pp. 23 figs.
Paraffin, 8 pp. 6 figs.
Pearl and Coral, 8 pp.
Perfumes, 10 pp.

Photography, 13 pp. 20 figs.
Pigments, 9 pp. 6 figs.
Pottery, 46 pp. 57 figs.
Printing and Engraving, 20 pp. 8 figs.
Rags.
Resinous and Gummy Substances, 75 pp. 16 figs.
Rope, 16 pp. 17 figs.
Salt, 31 pp. 23 figs.
Silk, 8 pp.
Silk Manufactures, 9 pp. 11 figs.
Skins, 5 pp.
Small Wares, 4 pp.
Soap and Glycerine, 39 pp. 45 figs.
Spices, 16 pp.
Sponge, 5 pp.
Starch, 9 pp. 10 figs.
Sugar, 155 pp. 134 figs.
Sulphur.
Tannin, 18 pp.
Tea, 12 pp.
Timber, 13 pp.
Varnish, 15 pp.
Vinegar, 5 pp.
Wax, 5 pp.
Wool, 2 pp.
Woollen Manufactures, 58 pp. 39 figs.

London: E. & F. N. SPON, 125, Strand.
New York: 35, Murray Street.

Crown 8vo, cloth, with illustrations, 5s.

WORKSHOP RECEIPTS,

FIRST SERIES.

By ERNEST SPON.

SYNOPSIS OF CONTENTS.

Bookbinding.
Bronzes and Bronzing.
Candles.
Cement.
Cleaning.
Colourwashing.
Concretes.
Dipping Acids.
Drawing Office Details.
Drying Oils.
Dynamite.
Electro-Metallurgy — (Cleaning, Dipping, Scratch-brushing, Batteries, Baths, and Deposits of every description).
Enamels.
Engraving on Wood, Copper, Gold, Silver, Steel, and Stone.
Etching and Aqua Tint.
Firework Making — (Rockets, Stars, Rains, Gerbes, Jets, Tourbillons, Candles, Fires, Lances, Lights, Wheels, Fire-balloons, and minor Fireworks).
Fluxes.
Foundry Mixtures.

Freezing.
Fulminates.
Furniture Creams, Oils, Polishes, Lacquers, and Pastes.
Gilding.
Glass Cutting, Cleaning, Frosting, Drilling, Darkening, Bending, Staining, and Painting.
Glass Making.
Glues.
Gold.
Graining.
Gums.
Gun Cotton.
Gunpowder.
Horn Working.
Indiarubber.
Japans, Japanning, and kindred processes.
Lacquers.
Lathing.
Lubricants.
Marble Working.
Matches.
Mortars.
Nitro-Glycerine.
Oils.

Paper.
Paper Hanging.
Painting in Oils, in Water Colours, as well as Fresco, House, Transparency, Sign, and Carriage Painting.
Photography.
Plastering.
Polishes.
Pottery — (Clays, Bodies, Glazes, Colours, Oils, Stains, Fluxes, Enamels, and Lustres).
Scouring.
Silvering.
Soap.
Solders.
Tanning.
Taxidermy.
Tempering Metals.
Treating Horn, Mother-o'-Pearl, and like substances.
Varnishes, Manufacture and Use of.
Veneering.
Washing.
Waterproofing.
Welding.

Besides Receipts relating to the lesser Technological matters and processes, such as the manufacture and use of Stencil Plates, Blacking, Crayons, Paste, Putty, Wax, Size, Alloys, Catgut, Tunbridge Ware, Picture Frame and Architectural Mouldings, Compos, Cameos, and others too numerous to mention.

London: E. & F. N. SPON, 125, Strand.
New York: 35, Murray Street.

Crown 8vo, cloth, 485 pages, with illustrations, 5s.

WORKSHOP RECEIPTS,
SECOND SERIES.

By ROBERT HALDANE.

SYNOPSIS OF CONTENTS.

Acidimetry and Alkalimetry.	Disinfectants.	Isinglass.
Albumen.	Dyeing, Staining, and Colouring.	Ivory substitutes.
Alcohol.	Essences.	Leather.
Alkaloids.	Extracts.	Luminous bodies.
Baking-powders.	Fireproofing.	Magnesia.
Bitters.	Gelatine, Glue, and Size.	Matches.
Bleaching.	Glycerine.	Paper.
Boiler Incrustations.	Gut.	Parchment.
Cements and Lutes.	Hydrogen peroxide.	Perchloric acid.
Cleansing.	Ink.	Potassium oxalate.
Confectionery.	Iodine.	Preserving.
Copying.	Iodoform.	

Pigments, Paint, and Painting: embracing the preparation of *Pigments*, including alumina lakes, blacks (animal, bone, Frankfort, ivory, **lamp**, sight, soot), blues (antimony, Antwerp, cobalt, cœruleum, Egyptian, manganate, Paris, Péligot, Prussian, smalt, ultramarine), browns (bistre, binau, sepia, sienna, umber, Vandyke), greens (baryta, Brighton, Brunswick, chrome, cobalt, Douglas, emerald, manganese, mitis, mountain, Prussian, **sap**, Scheele's, Schweinfurth, **titanium**, verdigris, zinc), **reds** (Brazilwood lake, carminated lake, carmine, Cassius purple, cobalt pink, cochineal lake, colcothar, Indian red, madder lake, red chalk, **red** lead, vermilion), whites **(alum,** baryta, Chinese, lead sulphate, white lead—by American, Dutch, French, German, Kremnitz, and Pattinson processes, precautions in making, and composition of commercial samples—whiting, Wilkinson's white, zinc white), yellows (chrome, gamboge, Naples, orpiment, realgar, yellow lakes); *Paint* (vehicles, testing oils, driers, grinding, storing, applying, priming, drying, filling, coats, brushes, surface, water-colours, removing smell, discoloration; miscellaneous paints—cement paint for carton-pierre, copper paint, gold paint, **iron** paint, lime paints, silicated paints, steatite paint, transparent paints, tungsten paints, window paint, zinc paints); *Painting* (general instructions, proportions of ingredients, measuring paint work; carriage painting—priming paint, best putty, finishing colour, cause of cracking, mixing the paints, oils, driers, and colours, varnishing, importance of washing vehicles, re-varnishing, **how** to dry paint; woodwork painting).

London: E. & F. N. SPON, 125, Strand.
New York: 35, Murray Street.

JUST PUBLISHED.

Crown 8vo, cloth, 480 pages, with 183 illustrations, 5s.

WORKSHOP RECEIPTS,
THIRD SERIES.

By C. G. WARNFORD LOCK.

Uniform with the First and Second Series.

SYNOPSIS OF CONTENTS.

Alloys.	Indium.	Rubidium.
Aluminium.	Iridium.	Ruthenium.
Antimony.	Iron and Steel.	Selenium.
Barium.	Lacquers and Lacquering.	Silver.
Beryllium.	Lanthanum.	Slag.
Bismuth.	Lead.	Sodium.
Cadmium.	Lithium.	Strontium.
Cæsium.	Lubricants.	Tantalum.
Calcium.	Magnesium.	Terbium.
Cerium.	Manganese.	Thallium.
Chromium.	Mercury.	Thorium.
Cobalt.	Mica.	Tin.
Copper.	Molybdenum.	Titanium.
Didymium.	Nickel.	Tungsten.
Electrics.	Niobium.	Uranium.
Enamels and Glazes.	Osmium.	Vanadium.
Erbium.	Palladium.	Yttrium.
Gallium.	Platinum.	Zinc.
Glass.	Potassium.	Zirconium.
Gold.	Rhodium.	

London: E. & F. N. SPON, 125, Strand.
New York: 35 Murray Street.

WORKSHOP RECEIPTS,

FOURTH SERIES,

DEVOTED MAINLY TO HANDICRAFTS & MECHANICAL SUBJECTS.

By C. G. WARNFORD LOCK.

250 Illustrations, with Complete Index, & a General Index to the Four Series.

Waterproofing — rubber goods, cuprammonium processes, miscellaneous preparations.

Packing and Storing articles of delicate odour or colour, of a deliquescent character, liable to ignition, apt to suffer from insects or damp, or easily broken.

Embalming and Preserving anatomical specimens.

Leather Polishes.

Cooling Air and Water, producing low temperatures, making ice, cooling syrups and solutions, and separating salts from liquors by refrigeration.

Pumps and Siphons, embracing every useful contrivance for raising and supplying water on a moderate scale, and moving corrosive, tenacious, and other liquids.

Desiccating — air- and water-ovens, and other appliances for drying natural and artificial products.

Distilling — water, tinctures, extracts, pharmaceutical preparations, essences, perfumes, and alcoholic liquids.

Emulsifying as required by pharmacists and photographers.

Evaporating — saline and other solutions, and liquids demanding special precautions.

Filtering — water, and solutions of various kinds.

Percolating and Macerating.

Electrotyping.

Stereotyping by both plaster and paper processes.

Bookbinding in all its details.

Straw Plaiting and the fabrication of baskets, matting, etc.

Musical Instruments — the preservation, tuning, and repair of pianos, harmoniums, musical boxes, etc.

Clock and Watch Mending — adapted for intelligent amateurs.

Photography — recent development in rapid processes, handy apparatus, numerous recipes for sensitizing and developing solutions, and applications to modern illustrative purposes.

London: E. & F. N. SPON, 125, Strand.

New York: 35, Murray Street.

JUST PUBLISHED.

In demy 8vo, cloth, 600 pages, and 1420 Illustrations, 6s.

SPONS'
MECHANIC'S OWN BOOK;
A MANUAL FOR HANDICRAFTSMEN AND AMATEURS.

CONTENTS.

Mechanical Drawing—Casting and Founding in Iron, Brass, Bronze, and other Alloys—Forging and Finishing Iron—Sheetmetal Working—Soldering, Brazing, and Burning—Carpentry and Joinery, embracing descriptions of some 400 Woods, over 200 Illustrations of Tools and their uses, Explanations (with Diagrams) of 116 joints and hinges, and Details of Construction of Workshop appliances, rough furniture, Garden and Yard Erections, and House Building—Cabinet-Making and Veneering—Carving and Fretcutting—Upholstery—Painting, Graining, and Marbling—Staining Furniture, Woods, Floors, and Fittings—Gilding, dead and bright, on various grounds—Polishing Marble, Metals, and Wood—Varnishing—Mechanical movements, illustrating contrivances for transmitting motion—Turning in Wood and Metals—Masonry, embracing Stonework, Brickwork, Terracotta, and Concrete—Roofing with Thatch, Tiles, Slates, Felt, Zinc, &c.—Glazing with and without putty, and lead glazing—Plastering and Whitewashing—Paper-hanging—Gas-fitting—Bell-hanging, ordinary and electric Systems—Lighting—Warming—Ventilating—Roads, Pavements, and Bridges—Hedges, Ditches, and Drains—Water Supply and Sanitation—Hints on House Construction suited to new countries.

London: E. & F. N. SPON, 125, Strand.
New York: 35, Murray Street.

www.ingramcontent.com/pod-product-compliance
Lightning Source LLC
Chambersburg PA
CBHW031846220426
43663CB00006B/516